T0327064

MACROEVOLUTIONARIES

MACROEVOLUTIONARIES

Reflections on Natural History,
Paleontology, and Stephen Jay Gould

BRUCE S. LIEBERMAN
AND NILES ELDREDGE

Columbia University Press
New York

Columbia University Press
Publishers Since 1893
New York   Chichester, West Sussex
cup.columbia.edu

Library of Congress Cataloging-in-Publication Data
Names: Lieberman, Bruce S., author. | Eldredge, Niles, author.
Title: Macroevolutionaries : reflections on natural history, paleontology,
    and Stephen Jay Gould / Bruce S. Lieberman and Niles Eldredge.
Description: New York : Columbia University Press, 2024. |
    Includes bibliographical references and index.
Identifiers: LCCN 2024009396 (print) | LCCN 2024009397 (ebook) |
    ISBN 9780231208109 (hardback) | ISBN 9780231557276 (ebook)
Subjects: LCSH: Evolution (Biology) | Natural history. | Paleontology. |
    Gould, Stephen Jay—Influence.
Classification: LCC QH366.2 .L54 2024 (print) | LCC QH366.2 (ebook) |
    DDC 560—dc23/eng/20240407
LC record available at https://lccn.loc.gov/2024009396
LC ebook record available at https://lccn.loc.gov/2024009397

Cover design: Henry Sene Yee
Cover images: American kestrel (Greg Hume); coyote (forestwonder); *Carcharodon
megalodon* (Bashford Dean); *Brontosaurus* (Bashford Dean); *Chesapecten jeffersonius*
(Berni Nonenmacher); *Nautilus pompilius* (Chris 73); *Eldredgeops milleri* (Daderot)

Contents

Preface

This book of essays examines how current topics in macroevolution and paleontology intersect with popular culture, philosophy, music, and the history of science, with an eye on the career and legacy of Stephen Jay Gould. Our essays are informed by the body of work we have produced during our lifelong analysis of the fossil record and our consideration of its relevance for understanding what George Gaylord Simpson once referred to as "the major features of evolution" (which we label more simply "macroevolution"). It is designed to foster thinking not just about the natural world around us but also on thinking itself. Part and parcel with this, our book seeks to convey what still inspires us after nearly a combined century dedicated to studying the history of life and macroevolution.

Each essay deals with one or more key topics in evolutionary biology and paleontology. Topics covered include the causes and consequences of mass extinctions, the nature and meaning of punctuated equilibria, and the relevance of paleontology to other scientific disciplines. Further, woven through these essays are thematic strands from music, history, and popular culture designed to show the links between science and society and, indeed, all realms of knowledge. In short, we seek to integrate the world of the intellectual with the more familiar human one and bridge the gap between noted philosopher Immanuel Kant and farcical Hollywood stuntman Super Dave Osborne, even though Kant never rode a motorcycle through a burning building and Dave never critiqued pure reason. As such, our essays are at times seriously reflective and at other times not overly serious.

One of our inspirations for this project was Steve Gould's life and legacy. We each knew him for decades. Steve and Niles attended graduate school together, and they collaborated on many papers; he also was Bruce's undergraduate advisor and mentored him thereafter. Steve perfected the art of the natural history essay, literally and figuratively, through his monthly pieces in *Natural History*. In homage, we seek to resurrect the natural history essay for a twenty-first-century audience during a time of revivified interest in the paleontological subject closest to his heart and ours: macroevolution.

Although we don't always ask or answer the toughest questions, such as "What did *Beavis and Butt-head* think of Elisabeth Vrba's turnover pulse hypothesis?" and "What city has the best natural history museum?," we do ask and answer some pretty tough ones, like "What did Elisabeth Vrba think of *Beavis and Butt-head*?" and "What city has the best pizza?" If you're looking for the answer to these and other questions more central to natural history, paleontology, and evolution, read on.

MACROEVOLUTIONARIES

CHAPTER 1

The Three Musketeers of Macroevolution

Does Anyone Get to Be D'Artagnan?

W e're going to tell you a story. It's about three paleontologists who spent their careers at Yale and Harvard and the American Museum of Natural History, respectively. There is also a fourth more junior paleontologist at the University of Kansas who aspired to join them. They lived, to put a twist on noted French author Alexandre Dumas, during a time when, at least in evolutionary biology and paleontology, panics were uncommon and (open) warfare was relatively rare.

The "Three Musketeers" may bring to mind a chocolatey, fluffy candy bar. Or the novel by Dumas, which paradoxically focuses mostly on a fourth individual, D'Artagnan, who joined the three senior principals. Or perhaps the first thing that comes to mind is a well-known Hollywood adaptation of that novel, starring Lana Turner and Gene Kelly, which depicted, tolerably well for its era, romances, fencing swashbucklers, and the nefarious Cardinal Richelieu. The last was played by that erstwhile exemplar of cinematic bad guys, Vincent Price.

But in certain rarefied circles, evolutionary and paleontological in scope, the phrase "the Three Musketeers" conjures up those three paleontologists our story began with. It was a phrase, in this context, invoked by Stephen Jay Gould—henceforth Steve—the member of the troika from Harvard, in the tribute page to his final publication, the 1,433-page tome *The Structure of Evolutionary Theory*. We quote:

> For Niles Eldredge and Elisabeth Vrba
> May we always be the Three Musketeers
> Prevailing with panache
> From our manic and scrappy inception at Dijon

To our nonsatanic and happy reception at Doomsday
All For One and One For All

Henceforth, these shall be treated as "The Three Musketeers of Macroevolution," in tribute to Steve's legacy. While it was originally only "according to Steve," given that one of us is a referenced principal (the now-retired museum curator) and the other a more junior paleontologist who studied with all of them and aspires to be the D'Artagnan to their Athos, Porthos, and Aramis, we think it fitting to echo Steve's phraseology. We posit that there might be Musketeers, not necessarily all self-styled, operating on some plane of paleontology in the arena of macroevolution, the subdiscipline dedicated to studying the major features in the history of life and understanding their relevance to evolutionary biology. And now we need to complete the trio by introducing the professor from Yale, Elisabeth Vrba. Oh, and Dijon: that's not only Bruce's second most favorite type of mustard (after honey), but it was also the French venue for a 1982 evolutionary biology conference when Steve first used the phrase "Three Musketeers" in reference to our trio. The Dijon conference focused on revisiting key issues in the field and establishing whether updates were needed to the dominant paradigm.

What brought these Musketeers together was a passion for evolution and a recognition that the study of paleontology would help advance the field. There was no literal foil, no Cardinal Richelieu, sitting in opposition to them. But some of their ideas, punctuated equilibria from one of us (Niles) and Steve and the turnover pulse (technically not until 1985; this was 1982) hypothesis from Elisabeth, did precipitate strong reactions from certain dissenting voices, who lacked the sangfroid and power of a Richelieu. (Of course, there was also lots of approbation and support for these ideas: in the end we believe the "good guys" won.) We will return to punctuated equilibria and turnover pulse in this book, but what about their exponents, Steve and Elisabeth?

WELCOME TO THE IVY LEAGUE

In light of these Ivy League professors, at Harvard and Yale no less, the uninitiated might expect authoritarian, top-down thinkers who held that their viewpoints should never be challenged and their opinions never be questioned. Imagine some stuffy professor. Bruce briefly got the chance to work for one of those early in his career, but he could only put up with it for about forty-eight hours before resigning. To set the scene, if you're on the way to becoming a professor but not quite there yet and living semester to semester as Bruce then was, picking up a small-scale teaching gig was something that you couldn't necessarily treat as beneath your dignity—not, that is, if you were concerned about being able to consistently pay for food and rent.

At one point during this stage of his career, Bruce was notified that an august Ivy league faculty member was looking for someone to coteach a class. This was sweetened by the suggestion that there would be a $2,800 paycheck associated with that activity. Not a great payout, but too large at the time for Bruce to ignore. So Bruce investigated further. Step 1 was to talk to folks in the department administering the class, given that they would be the faculty and staff who would approve any paycheck for the gig. At that meeting they asserted that the gig of coinstructor at this university might pay $2,800, but that was a maximum and subject to scaling based on enrollment levels; they in turn told Bruce to be sure to show up to the classroom where the faculty member was teaching, ten minutes before showtime, and be ready to get in on some dramatic pedagogy. This news was not as good as the original news had been, but there was still the hope that the job of coinstructor at a grand university could bring some more shine to Bruce's CV.

The room where the class was held was a grim affair: classic 1950s modernist architecture, no windows, tucked into the back hallways of the building. Bruce was somewhat dismayed to see that there were only three students (coincidentally equal to the number of Musketeers) in the class, indicating that the full $2,800 could be slipping from his grasp, but he became even more dismayed to find that coteaching with this faculty member meant principally placing slides into a carousel and running the projector while the professor droned on and on about subjects that were both rather inscrutable and had nothing to do with paleontology, Bruce's subject area.

The teaching gig took an even sharper turn for the worse at the conclusion of class, when the professor indicated Bruce should carry the slides and projector back to his office. There, upon arrival, he duly informed Bruce that he should make the coffee, and please be sure to add two lumps of sugar. Truthfully, Bruce was so flummoxed at this point that his only action was to brew up some coffee with the requested number of lumps of sugar. Then, mumbling something to the effect of "happy to provide additional teaching 'assistance,' if the matter of receiving necessary fiduciary compensation was approved by the department," Bruce was able to beat a hasty retreat.

But he recognized that for his sanity to be maintained, and in order to be able to simply look himself in the mirror, the $2,800 that had originally been proffered might not be enough. The next day Bruce steeled himself for his visit to the department administration, when pen would supposedly be set to paper, and salary and coinstructorship would be granted. However, this occasion was even less salubrious than the previous ones of this storyline, concluding with a remark that Bruce to this day remembers distinctly: "Due to limited enrollment, alas we can only pay you $800 to continue on as a TA." The title of "coinstructor" had been removed, and Bruce had been subject to the classic used car lot bait and switch. "However, it is our understanding that this professor is such an amazing scientist that you really should be willing to do this for free," the bureaucrat continued. Well, Bruce was a bit taken aback by this, to be sure, even though he replied as politely as possible, "Alas, at this time I need to decline your generous offer." But at least now he had a way out of this crappy gig as quasi-barista.

Back to the professor's office, where thankfully the coffee had already been made, and Bruce, trying to be as politic as possible, said he would have loved to have had the opportunity to teach together, but the amount of money the university would be providing would not be sufficient to satisfy basic needs and wants. The professor gave a noncommittal nod of basic understanding, though he went on to remark, "You really should be willing to do this for free." And that's why Bruce to this day appreciates the fact that Steve and Elisabeth weren't overly hierarchical, bossy, and pedantic, especially by Ivy League professor standards.

One of the reasons that both Steve and Elisabeth adopted a posture somewhat different from your standard Ivy League professor may at least partly have to do with the fact that neither came from a privileged background, and both came from groups that were, at the time they were working (and still to some extent today), quite underrepresented in the field in general and their respective departments in particular. As far as we can determine, Elisabeth was the first female tenured faculty member in the Department of Geology & Geophysics at Yale, and Steve was the first Jewish professor in the Department of Geological Sciences at Harvard. While to non-Jews that may not seem like much of a stretch, never criticize someone or their feelings until you've walked a mile in their shoes. At the very least, by that point you'll be a mile away and you'll have their shoes, as humorist Jack Handey so wryly noted, thus there's not much they can do about that criticism if they don't happen to like it. We're not saying that Steve or Elisabeth experienced *open* hostility from their fellow professors because of their background nor more hostility than others have or still do experience today. However, certain department customs and approaches would have been highly at odds with what a New Yorker who attended public schools, as Steve did, would generally experience. Some departments had decidedly distinctive customs, such as afternoon sherry-sipping clad in jacket and tie, which at least to us seems highly incongruous with a modern perspective on science in the way that a cup of coffee or even a beer or martini would not.

We do think that some of the lingering resentment directed at Steve as he became more and more successful, and toward Elisabeth as she did, had to do with the fact that geology departments can be notoriously stuffy by academic standards. Specifically in the case of Elisabeth, the resentment generally took the form of trying to ignore her and her accomplishments, including by trying to avoid citing her work. But other times it was more confrontational. Bruce remembers one time giving a talk at a regional public university in a small northeastern state and having a faculty member literally start yelling at him because he dared to have studied with Elisabeth, given that she had actually posited that climate change played a role in shaping the evolution of the human lineage. (This is her turnover pulse hypothesis, which we shall describe in greater detail later.) Because it was for a job interview Bruce mostly had to grin and bear it, but a response of "Whatever, man" would have most assuredly been in order. Perhaps regrettably, at least when it came to his potential status as a D'Artagnan, Bruce didn't stand up for Elisabeth's scientific honor nor demand a duel on the spot. Or what was overheard by Bruce's wife at a conference by a woman who went on and on unprompted in a group setting trying to describe how horrendous Steve was as a person and a scientist (even though he wasn't at the conference and she had never met him). And what some may have interpreted her criticism amounted to was that she felt that Steve was "a know-it-all Jew." Come to think of it, if she had known the term, she might have said that he had "a lot of chutzpah," which she probably would have pronounced "shoots pah."

Some people in academia, and elsewhere for that matter, try to bully or put down people whom they see as easy targets. That's because they may be viewed as outsiders and thereby lacking the support that might make one otherwise think twice about such bullying.

But it's more than just this. It was partly that Elisabeth and Steve had the gumption to "put themselves out there": to opine about science and theories that maybe some thought paleontologists shouldn't have the temerity to say anything about—for instance, to consider organizing principles of evolution or the fact that competition among life forms is not what generally causes extinction. People love to criticize, especially the folks who "do," and Elisabeth and Steve were definitely "doers," which is one of the reasons they ended up at Yale and Harvard, respectively. As former president and Nobel Peace Prize winner Teddy Roosevelt—a great thinker, albeit a flawed one, but one who did help found the American Museum and, among the best ideas this country produced, the National Park System—remarked, "It is not the critic who counts; not the man who points out how the strong man stumbles, or where the doer of deeds could have done them better. The credit belongs to the man who is actually in the arena, whose face is marred by dust and sweat and blood; who strives valiantly; who errs, who comes short again and again, because there is no effort without error and shortcoming." (We'd like to think that Roosevelt, being a wise fellow, would have said "human" or "person" or somesuch to allow women a place in the discourse.) Because, while science is not truly a war, intellectual and emotional skirmishes definitely do occur among thinkers in the arena of ideas. As Niles and Steve recognized, getting scientists to accept their new idea about the tempo and mode of evolution, punctuated equilibria, was not just about accumulating numerous facts, it was about getting them to think in a new and different way. The same was true of turnover pulse, which essentially took punctuated equilibria, which focused on single lineages, and scaled it up to the level of regional biotas possessing many species.

Another issue is that both Elisabeth and Steve put up protective walls. For instance, Elisabeth was very defensive of her time, partly because she was asked to do so much service as the only woman in Yale's Department of Geology & Geophysics. If she said "yes" to everything and everyone she'd quickly be overwhelmed with tasks that were not really beneficial to her career. And Steve put on an air that was intimidating to strangers. This was misunderstood as "snooty" behavior. It was instead akin to the pose one has to adopt as a resident of New York City: you learn not to have a continually smiling and affable demeanor or to greet every stranger with a welcoming "hello" and a firm look in the eye. If you appear too friendly, goodness knows what type of person is going to come up and start talking to you on the street or in the subway, conversations that tend not to be "beneficial to one's career." Further, kindness is often mistaken for weakness. Of course, Steve had to adopt such a pose on steroids. Let's face it, a lot of people are "Star F—ers" and just want to get some association with and monopolize the time of those that are widely recognized. Although it might not be the literal equivalent of being accosted on a subway platform, one cannot engage with everyone that desires engagement, especially if they're engaging for their own purposes. Basically, Steve and Elisabeth were held to a biased double standard that was likely rooted in their backgrounds and their status as "outsiders" who did not belong to the prevailing majority.

In any event, intellectual skirmishes were why Steve invoked the term "Musketeer." He wanted to view Niles and Elisabeth as having his back in an intellectual space. To join him in a fight against scientific intransigence by what he saw as many mini-Richelieus. It's also how he wished they'd view him. He hoped they'd be willing to go into an intellectual fencing match with others who diverged from or were downright skeptical about the ideas they agreed upon—for instance, that punctuated equilibria is valid, or that natural selection operates on many levels such as the gene, the organism, and the species. But Steve, as with any person, even a Musketeer, often went solo, promoting, modifying, and sometimes getting more than his share of the credit and attention for these ideas.

Key parts of the idea in the 1972 scientific paper that presented punctuated equilibria to the world appeared in the 1971 paper that Niles published alone. As we shall detail in chapter 12, Elisabeth also was the originator of the key tenets of exaptation, which is often credited solely to Steve. In fact, there were aspects of evolutionary theory where Niles and Elisabeth diverged from Steve and had more in common with each other, or maybe even someone Steve would count as a mini-Richelieu.

There can be advantages to being a part of a group of Musketeers. But it also can set you up as a target to outsiders or make you too unsuspecting when it comes to members of the group itself. As Trinidad Silva Jr.'s character put it so wisely near the end of the 1988 movie *Colors*, "You screwed up now, man . . . You joined a gang." Never join a gang, especially in academia. The level of loyalty is limited, the focus on self typically prodigious, and attempts at mutual back scratching can often go unreciprocated.

Niles and Elisabeth never felt that Steve did badly by them, but there were certainly instances of hurt feelings; Bruce always wanted to try and keep the filial connection to all of them in the hope that the band would never truly split up. But science is not an actual life-and-death struggle, so you don't need a real band of musketeers. (Friendly reviewers of papers and grants can be highly beneficial, though!)

Bruce was certainly of a younger generation, as D'Artagnan was, but he never took part, at least with them, in the intellectual skirmishes belonging to the world of Niles, Steve, and Elisabeth when they were trying to get the nonpaleontological field of evolutionary biology to pay attention to their ideas. These episodes occurred before he arrived on the scene. Maybe if he had gone to graduate school ten years earlier he could have been the true D'Artagnan to their Three Musketeers. But he arrived on the scene too late and after the team had split up, as it were. They had all moved on and were doing their own thing. Steve wrote those last words in 2002 because he was feeling nostalgic upon completing his last book, and because he knew he was dying. Death can be a time for nostalgia. It was also Steve's last, best chance to publicly show appreciation and respect. So it was a cool nod to the past.

HELLO, ELISABETH

"Do you understand what I'm saying?" Elisabeth asked with a trace of insistence (figure 1.1). "Uh, yeah," Bruce said, staring at the taxidermic mount of a head of an impala, with its charismatic but not exceptionally ornate horns, mounted on the deep blue cinder block walls of her office. Bruce had walked over from his office down the hall, its cinderblock walls painted a shade of dark red.

Elisabeth would ask Bruce if he understood what she was saying several times during that and many other conversations. She had a habit of saying it. "Do you know what I'm saying?" is seemingly a very modern and hip invocation in the 2020s, but this was in the mid-1990s, so Elisabeth was well ahead of her time on this, as she was on so many matters scientific.

Bruce was wondering why she said it so often. Did she truly think he didn't understand, or was it just a figure of speech or a rhetorical device? He was hoping it was the latter. Elisabeth was very knowledgeable and intellectually formidable, so perhaps she was used to people not necessarily initially grasping what she was saying. At their first meeting Elisabeth was trying to explain how she had recognized that climate change triggered the evolution and extinction of multiple

1.1 From left to right: Niles Eldredge, Elisabeth Vrba, and Stephen Jay Gould in the latter half of the 1980s. Photo by George Vrba, used with permission.

species in a region, her turnover pulse hypothesis. She was at this particular instant focusing on how the evolutionary changes were tied to changes in the rate that these organisms developed into adulthood and sexual maturity. And she was using several quantitative approaches to test this. Bruce was in New Haven, the home of Yale and its collegiate fight song "Boolah Boolah." (The greatest version of that song was performed by the legendary Duke Ellington in the early 1970s; do yourself a favor and give it a listen.) Bruce was there working on a postdoctoral fellowship with Elisabeth. A postdoctoral fellowship represents the "sojourn" between graduate school and a faculty job, when aspirational young scientists work to hone their craft or expand their toolkit. (More of this in chapter 13.) It's like being promoted to the triple A minor leagues in professional baseball or the development or G-league in professional basketball, when you're almost but not quite in the big leagues.

Bruce had told Elisabeth that he wanted to continue some of his molecular work, begun as a graduate student at the AMNH, via a stay in the molecular lab of Yale faculty Jeff Powell and Gisella Caccone, to further test and study some of her ideas on evolution. This was the one Elisabeth and Bruce were trying to formulate. He was hoping it would work, and he had chosen freshwater clams from the rivers, streams, and lakes of the middle part of North America as the study organisms. Elisabeth, while her specialty was modern and fossil (from the last five million years or so) tropical mammals from Africa, was always broad-minded enough never to restrict herself to one group or time. Thus, thankfully, she assented to the molluscan focus. In any event, Bruce was stoked to get to work with his third Musketeer in a row, though he'll save his encounter with Niles for another venue.

Niles's first "meeting" with Elisabeth occurred via a manuscript sent by transatlantic mail in 1980—boat rate, and thus over the transom. She was introducing herself and soliciting comments. (Simultaneously, she had sent a copy to Steve.) Niles immediately recognized the manuscript's significance: it dealt with many things, including the causes of trends and how rates of speciation and extinction could be regulated by the ecological properties of organisms, the latter of which she termed the "effect" hypothesis. Niles was very excited to have encountered such a bold and insightful thinker. Elisabeth's timing with the manuscript was impeccable, even if the mode of postal delivery lacked celerity. There was an upcoming meeting on macroevolution at the Field Museum of Natural History in Chicago, a conference that is today remembered as historic and significant. Niles was one of the conference organizers, along with Jeff Levinton, Joel Cracraft, and David Raup, and they made sure that Elisabeth was invited. There, at last, Niles met her in person. *Science* magazine reporter Roger Lewin provided a lengthy writeup of the conference in that prestigious journal, featuring quotes from Elisabeth and others. Significantly, Lewin's writeup also featured several figures from Elisabeth's manuscript, which by then had been published. Talk about a coming out party: for a scientist to have their figures featured in an article in *Science* was a high accomplishment indeed, but it was the first of many for Elisabeth.

HELLO, STEVE

Bruce's first meeting with Steve was certainly an experience (figure 1.2). It was his first encounter with any of the Musketeers, though at the time he had no inkling that there were three Musketeers, at least beyond the realm of chocolate, novels, or Hollywood. Cut to the preamble. Bruce was enrolled as an undergraduate in Steve's modestly titled class "The History of Earth and of Life." Seriously, who expects to be able to cover one of those topics, not to mention both, in a single semester, or even a single lifetime? But Steve seemed to think he could. Steve was a great lecturer, no doubt about it, and this class was legendary. Indeed, it was in high demand among Harvard undergraduates for several decades. This demand arose because of Steve's brilliant elocution skills and the nature of the subject matter; general interest in paleontology runs high, but he was also a famous, perhaps the most famous, professor on campus. Many folks were there simply because they wanted to be able to say they took a class from "the great Stephen Jay Gould." Bragging rights for some cocktail party twenty years in the future was something that seemed to motivate certain students on campus. And for sure the course had the added benefit of also fulfilling a requirement of the university's dreaded "core curriculum." The class was offered at the same time as another class that happened to meet the same core curriculum requirement, taught by Steve's bitter nemesis E. O. Wilson. However, at the time we undergraduates were only dimly aware of disputes involving faculty fiefdoms. But Bruce was taking "The History of Earth AND of Life" instead of Wilson's class, and it ultimately seems to have been a good career choice, because that course was at least the proximate reason he decided to become a paleontologist. That in turn led to the introduction to his eventual PhD advisor and coauthor here, Niles; and that in turn led to a postdoctoral position with Elisabeth.

Bruce's experience with each of the Three Musketeers thus had its birth in the seemingly simple decision to take that large class, which enrolled three hundred-plus students. It truly came to pass because it was offered at a day and time (not too early) that worked for Bruce and his two roommates, Mark and Ron (another trio), so they all could take it together. Probably E. O. Wilson would have argued that our decision to co-enroll was instinctual and derived from our forebears' experience on the African savannah millions of years ago, when it would have been dangerous for a solitary primate to venture out on their ownsome when large, deadly groups of conspecific primates might be encountered. That is, Wilson might have posited a sociobiological, Darwinian basis for this choice. But the genesis of the decision was far more mundane and in Gould-ian parlance "contingent," a concept we shall return to, especially in chapter 4. Bruce, Mark, and Ron were all interested in the subject matter, and further, Mark had made the shocking decision of becoming a geology major. "What even was that?" Bruce and Ron wondered, and "For goodness sakes why?" It so

1.2 Stephen Jay Gould in his office at the Museum of Comparative Zoology, Harvard University, sometime in the 1990s. Image courtesy of the Department of Special Collections (DSC), Stanford University Libraries (SUL), and Tim Noakes, Head of Public Services, DSC/SUL. Used with permission; photographer unknown.

happened that the class provided within-major credit on top of core curriculum credit, a win-win: Mark's realization of this fact indicated that he didn't go to Harvard for nothing. (Little did Bruce know that he was also soon to become a geology major.) Further, to rebut the Darwinian argument, we weren't worried about being slaughtered or worse by our coregistrants in Harvard Yard's joyless Science Center building. Instead, it is a well-known fact that any class was always better with friends, because it's easier to stay au courant with course material if you've got someone you can study with or get notes from if you happened to sleep through a class. Sociobiology is a load of claptrap when it comes to auguring human behavior, however well it might work in certain social insect societies. Bruce can't imagine ever pursuing a career in sociobiology, which is what E. O. Wilson lectured about, so he is grateful for the decision by his roommates, especially Mark, that put him in there at that point in his life when he was intellectually adrift, searching for a college major. That was a key contingent moment in Bruce's career.

Bruce's initial impression of Steve focused on the physical. Steve was short and, we hate to say it, kind of schleppy. He resembled folk/pop singer Paul Simon in the early stages of his solo career, with a mustache and a little extra weight. Steve even ambled into class wearing the same winter jacket that Simon sported (albeit ripped and with the fur partly detached) on the cover of his self-titled first solo album. But when he started talking, it was clear that physical appearances were deceiving: the dude was bloody brilliant. Edie Brickell, Paul Simon's wife and a great songwriter in her own right, posited that "philosophy is the talk on a cereal box, religion is the smile on a dog." While we do think that "a walk on some slippery rocks" could have worked tolerably well for part of that first verse, even so, Steve came to lecture throwing out proverbial cereal boxes and commenting on smiling dogs left and right while at the same time providing incisive commentary on paleontology.

Bruce, his roommates, and all of his classmates held that no one constrained by the dictates of a less than genius intellect would be showing pictures from treatises by seventeenth-century creationists, concomitantly positing that no less a personage than the great Charles Darwin had made factual errors in judgment while later showing a picture of the ceiling of St. Mark's Cathedral in Venice, Italy, not too long after providing an exegesis on small Cambrian fossils from the Rocky Mountains of Canada and their misinterpretation by the former director of the Smithsonian Institution. Which was what Steve did throughout the course of the semester. How could he grasp all these disparate subjects?

And the recognition of that brilliance was not solely confined to students in the classroom. Steve arguably became in his lifetime the best-known scientist in America. He won a MacArthur Genius award. He was a member of the prestigious National Academy of Sciences. He was president of the American Association for the Advancement of Science. He published dozens of bestselling books, along with hundreds of scientific papers. He had a large coterie of successful graduate students today placed in numerous faculty positions. He was even featured as a character on *The Simpsons*. It's a career to be envied.

Yet today, if you visit Harvard University's website on the history of his erstwhile scholarly home, the Museum of Comparative Zoology, how do they describe him? "Steve Gould, whose monographs on the evolution of land snails, and whose general writings on macroevolutionary theory have provoked much discussion and controversy in evolutionary studies." Sheesh! Talk about damning with faint praise. By contrast, they describe Bernie Kummel as doing "more than any other paleontologist to establish the empirical basis of patterns in faunal decline and recovery around events of mass extinction" and Harry Whittington as "the world's leading trilobite taxonomist and promulgator of the distinguished research program that fundamentally reinterpreted the Burgess Shale and the meaning of the Cambrian Explosion in evolutionary terms." No knock on Kummel or Whittington, but it was Steve who probably did more than anyone else to advance the study of mass extinctions, the Burgess Shale, and the Cambrian Explosion, all topics we will visit in this book. Well, we introduce this merely to prove the validity of the motto "*sic transit gloria*." If you're famous and highly regarded, you better not die. Or when you do, you better have someone (dare we say, Musketeers) who will do battle for you. Otherwise your reputation will be tarnished, maybe even ground into dust by those seeking to use the mountain of work you created to climb ever higher, without giving appropriate credit. Recall that *Ozymandias* poem by Percy Bysshe Shelley. Or the Kinks' *Celluloid Heroes*: "For those that are successful, be always on your guard, for success walks hand in hand with failure." Indeed, this is true not just on the Hollywood Boulevard, but on the grand promenade of science and life.

Meeting Steve in person for that first time happened during the course's office hours, when Bruce had made an appointment to discuss becoming a major in paleontology via Harvard's Department of Geological Sciences. The lead-up was a slightly nerve-wracking experience. The nerves were wasted energy, as they almost always are. At those first office hours Bruce was confined to the margins of a circle comprising a substantial number of sycophants who had showed up as well, not because they were interested in the subject matter or in becoming paleontologists but rather because they wanted to be able to say not only that they took a class with the great Stephen Jay Gould but also that they spoke with him. Thus, for practical purposes, the first meeting had to be postponed a week, and thus something to obsess on and get nervous about for an additional seven days. Why be nervous? What do you say to someone like Steve if you're a college sophomore, and not particularly even a wise fool, as that word means in Greek? There's no hope of impressing them; you hope only to avoid doing something deeply embarrassing. Thankfully, Bruce managed to avoid the latter while obviously eschewing the former. Bruce still insists that one of the things he did that most impressed Steve over the course of their decades of interaction was using the word "obdurate" in a paper we read and discussed during a weekly lab meeting also attended by some of Steve's graduate students, including Linda Ivany, now on the faculty at Syracuse University. Bruce used the word only because he was intrigued/inspired when he heard Richard Lewontin say "obdurately monomorphic" in a seminar years before as he bemoaned the lack of genetic variation in certain species. Go figure! Notably Lewontin, whom we will discuss in chapter 6, was Steve's informal mentor, and Steve idolized him (and thankfully he happened not to attend that seminar).

Niles first encountered Steve in 1963. Prior to paleontology, Niles's scientific interests ran more to the anthropological, and he was especially intrigued by distinct cultures and how place and environment affected their development. (His evolutionary ideas continued to have such a focus, emphasizing how the physical environment triggered speciation, extinction, and ecological change, as we shall document throughout our book.) In the summer of 1963, after finishing up his sophomore year of college, Niles had been conducting anthropological research (a polite term for bothering people) in the beautiful country of Brazil, using a broken, baby-talk form of Portuguese. At this time Niles's girlfriend and future wife, Michelle, had split for her ancestral California home. The early anthropological research convinced Niles that evolution and fossils were his true scientific passion, while his interactions with Michelle convinced him she was his true, overarching passion. He flew out to Berkeley around Christmas time to persuade Michelle to marry him, and thankfully the trip was a success. The official ceremony took place in her parents' garden in June 1964.

In the fall of 1963 Niles was beginning to hang with the new graduate students in Columbia University's Department of Geological Sciences. These students were principally housed in Schermerhorn Hall on Columbia's Morningside Heights campus in New York City, which still lines Amsterdam Avenue hard by 118th Street. The building was the home not only to geology but also zoology, anthropology, fine arts, and psychology (urban campuses can get pretty packed). It was also famous as the erstwhile scientific home of Thomas Hunt Morgan early in the twentieth century, when he and his lab group were conducting their epochal studies on fruit flies and the principles of genetic heredity. Other famous work conducted there included research associated with the Manhattan Project.

A distinctive aspect of the building was the inscription above the door: "For the advancement of natural science. Speak to the earth and it shall teach thee." This reference to the Book of Job might seem anachronistic in a modern scientific context, and it continues to vex ichthyologists who justifiably feel deflated that the succeeding biblical verse, "And the fish of the sea shall declare unto thee," was left out. However, we feel the phrase lends the building a certain grandeur, although we are admittedly biased given that some of our key, early experiences as paleontology students took place in that building. That was where Niles took his first paleontology course, which included a lab exam on snails (gastropods, or better, ghastly-pods) that he shall never forget so long as he lives, given that was where he was and what he was doing when John F. Kennedy was shot.

Steve, a new graduate student in geology, and thus an upperclassman relative to Niles, who was still in college, had grown frustrated that there was little to no evolution covered in Columbia's paleontology curriculum of the era. This is thankfully a condition much less prevalent anywhere today, including at Columbia, in large part due to the discovery of punctuated equilibria and happenings like that 1980 macroevolution conference in Chicago. However, it was very much the case *everywhere*, not just at Columbia, up through at least the mid-1980s. Steve's frustration thankfully precipitated action. He formed a student seminar on the subject, and Niles was grateful to be included. The rest, as they say, is history.

SOME PARTING THOUGHTS

While Steve was certainly not a great competitive fencer in the physical realm, à la seventeenth-century French musketeers, when it came to verbal sparring or duels in print he was very hard to defeat. His logic was generally impeccable and impervious to attack, a trait that he and Elisabeth shared. Steve's knowledge was voluminous across many areas, so he could draw on many subjects to make his point. Elisabeth was more of a biologist than Steve was. She also had extensive training in anthropology, which Steve lacked. Steve's rhetorical and literary skills were legendary, garnered via experience honed over many decades, just like the fighting tricks of the trade that Athos, Porthos, Aramis, and D'Artagnan possessed. Few successfully challenged Steve in person or in print. Elisabeth was less of a rhetorician, but so was just about everyone, and she made fewer personal appearances at conferences than Steve. Steve thus managed to interact with more people than Elisabeth did, but those who did interact with her knew how truly brilliant she was.

The only thing that could ultimately vanquish Steve and his wit was the final curtain call we all must answer, death. Everyone, especially paleontologists, knows that's a fate that awaits us all. But just like Dumas's work and his characters, which live on after he expired, so too have Steve's work and his ideas.

There are always multifarious reasons for one's success, but generally a key attribute all successful people possess is a great degree of talent and tenacity. Steve and Elisabeth possessed each of these qualities in abundance, which spurred them on from humble beginnings to Ivy League professorships. That's why people continue to debate and argue about them and their work long after they retired. However, with these macroevolutionary Musketeers, unlike Dumas's, it never really was "all for one, and one for all." But maybe that's to be expected. The work is fiction, though partly inspired by real events, and concocted by a collective of writers whom Dumas managed. Rather, the macroevolutionary Musketeers all had a general affinity for the same types of interests, ideas, and ways of thinking. In the case of Niles and Steve this affinity can be partly explained by a shared common ancestor in their academic lineage, graduate advisor Norman Newell, along with the fact that they both grew up in New York City or its metropolitan surroundings and attended Columbia University for graduate school during the same general era. Thus, some of these interests were developed in an ancestral-descendant framework intellectually. By contrast, Elisabeth developed hers independently in an entirely different milieu, having been raised in Namibia (German Southwest Africa when she arrived) and trained as a physical anthropologist and vertebrate paleontologist with an interest in hominid origins in South Africa.

But the end of Dumas's story somewhat resembles the end of our own. Alas, the real, fictional Three Musketeers largely split up or drifted apart. One, Porthos, left the service (that's Elisabeth). The other, Aramis, disappeared entirely (that's Steve). Athos (Niles) took a job working for D'Artagnan, who had received a big-time promotion. But Bruce is not D'Artagnan, so even though he is the first author of this book, Niles is most definitely not working for him or anyone else. Instead, he's enjoying retirement and writing, though Bruce rushes to remind him that there could be worse people to work for . . . And Bruce is still hoping for that big-time promotion. But even if he doesn't get it, at the end of the day, he's very glad he got the chance to know and work with those honest to goodness, flesh and blood Musketeers.

CHAPTER 2

Asleep at the Switch

Paleontological Life Lessons, Stasis, and the Genius of Yogi Berra

There are certain things most of us learned or should have learned as children, such as obeying the admonition not to run with scissors. If we consider the narrower disciplinary purview of paleontology, there are also certain equivalent life lessons, or better "history of life" lessons, that most of us paleontologists learned. One of the key examples we'll focus on in this chapter involves the nature of evolutionary change within species. What pertinent history of life lessons have paleontologists revealed? The big one is that if you search for evidence of persistent, directional evolutionary change within species lineages in the fossil record, you don't find any. You literally and figuratively slam into a stone wall. Regarding the literal part, this happens because you are most likely sampling up and down a cliff face at an outcrop somewhere—and you slam into *that* kind of stone wall. So don't run with scissors, and don't run around outcrops.

Once you cut it out with the running and start poking around the outcrop, using the trusty hammer you brought along to liberate a few fossils from the various rock layers, what do you see? First, we feel compelled to quote one of our favorite philosophers here, Yogi Berra, because this quote conveys such an important life and history of life lesson: "You can see a lot just by looking."

Of course, you can't "see" anything without looking. Further, if you're a scientist, what you should do is look around for patterns, try to synthesize them, and use them to test and refine hypotheses. When you look at specimens of a species from the bottom to the top of your rock wall, you'll find that the fossils of the lineage you are focused on don't seem to change very much. They all look pretty much the same: that is, they display *stasis*. The first and still classic example—we

2.1 The head of the trilobite *Eldredgeops rana*.

can even call it the paradigmatic example of this pattern and the associated phenomenon of *punctuated equilibria*—involves Devonian trilobites of the genus *Phacops* (figure 2.1), which is now due to a fit of taxonomic pique sometimes called *Eldredgeops*.

If you haven't given up at this point and still want to follow Yogi's maxim and keep on looking, you shrug your shoulders, get into your car, and drive to the next promising outcrop. Let's say the rocks here are said to be a bit older than the rocks you just banged up against and banged on in the first quarry. If we look for our species and find it, the fossils will look pretty much the same as the ones you saw in that first place. There might be a few very minor differences, equivalent to the types of differences between closely similar populations of modern species that varied slightly across geographic space. This process of tracking down your chosen fossil species lineage and looking for change within it could go on and on till you run out of patience or gas, or until you come upon an outcrop that's too far back in time, that is built up of layers of rock deposited before your species evolved, or you reach an outcrop that contains rocks that are too young and contains layers of rock deposited after your species went extinct. Given that on average marine invertebrate fossil species, such as our trilobite, persist for 3–10 million years, that's a lot of time without a lot of net within-lineage evolutionary change. What does this mean about the nature of the evolutionary process? What is the history of the concept of stasis? What causes it? These questions are of key relevance to macroevolution, and ones that very much motivated the scientific career of Stephen Jay Gould.

STASIS AS A PATTERN

Now, the paucity of *net* change in morphology could actually fit several different patterns: (1) obdurate monomorphism, with mean or median species morphology not diverging statistically from its initial morphology at any point throughout its history; (2) waggling with concomitant wiggling in morphology, involving repeated cyclical changes in mean or median species morphology (figure 2.2); and (3) meandering morphological changes in mean or median morphology that lead nowhere that could be characterized in a gestalt sense as a directionless random walk (figure 2.3).

All these patterns are compatible with stasis as originally defined in the first 1972 paper on punctuated equilibria by Niles and Steve, and as characterized in many subsequent papers and books they authored, together (including with Bruce) or in collaborations with others. It seems like some folks are confused and don't get the fact that as long as the mean or median morphology of a species from its first and last occurrence does not differ, then what is observed is compatible with stasis as originally defined. This holds true even if there is some sort of random walk in morphology that occurs during the history of the species. We can say this because we've been there and produced the original box score, if you will. Further, we think Yogi Berra would agree that before other scientists try to say that some result proves or rejects stasis they really need to read the original, to check the official record and figure out what the original authors actually said. You really can see a lot just by looking. Steve, being a Yankees fan, would surely have agreed with our proverbial tip of the cap to Yogi.

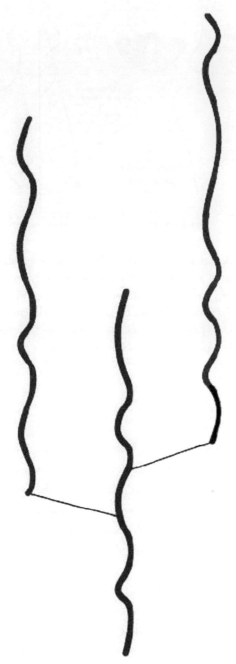

2.2 What typically happens to the morphology of species over long stretches of time: oscillations that don't go far and yield little net change. Time is reflected on the vertical axis and morphology is reflected on the horizontal axis. The central squiggly line represents an ancestral species that at one point during its history speciated, with speciation represented by the thin diagonal line. It gave rise to two descendant species, depicted by the left and right squiggly lines. From N. Eldredge et al., "The Dynamics of Evolutionary Stasis," *Paleobiology* 31 (2005): 133-45, copyright © Paleontological Society, used with permission.

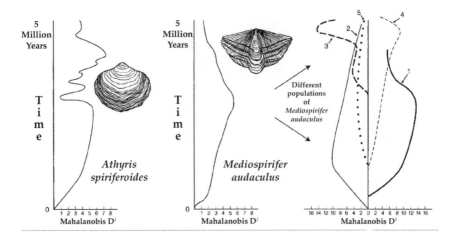

2.3 Left and center, schematic diagram showing patterns of change in the morphology of two brachiopod species, *Mediospirifer audaculus* and *Athyris spiriferoides* over the course of roughly five million years. On the right, the patterns of morphological change within populations of *Mediospirifer audaculus* broken down into the component environments they occurred in. From N. Eldredge et al., "The Dynamics of Evolutionary Stasis," *Paleobiology* 31 (2005): 133–45, copyright © Paleontological Society, used with permission.

STASIS AS A PROCESS

The process that produced the stasis we'd see in a lineage with morphology that oscillates cyclically through time is not the same as the process that would produce an obdurately monomorphic lineage, nor is it the same as the process that causes morphology to drift with evanescent directions in a random walk. Thus, there are likely several processes that cause stasis, and the process or processes generating stasis have been much considered and debated by several scientists, including Niles and Bruce along with Steve.

One of the ideas that Niles and Steve originally put forward in their 1972 paper introducing "punctuated equilibria" was that there could be certain constraints in the way an organism develops from a fertilized egg to an adult that would serve to make evolutionary change harder to occur. This was an idea that held special significance for Steve, who spent important parts of his career focusing on how development influences evolution. He produced many great works associated with this over the years, but inarguably the most profound and influential of these works is his epochal 1977 book *Ontogeny and Phylogeny*,[1] one of the key scholarly publications in this area. Quite frankly, Steve does not get much credit anymore for this accomplishment in this area, except for maybe a desultory citation or two, yet he richly deserves it; verily, he should be considered one of the founding figures of the field of evolutionary developmental biology. Today research in this area, sometimes called evo-devo, is providing fundamental insights to many important questions in macroevolution. However, later in his career, while still envisioning development as a key aspect of evolution, Steve acknowledged that developmental constraints might not be one of the key factors that produces stasis. This is because humans have been able to produce tremendous shifts in morphology in various domesticated organisms over rapid time scales via artificial breeding and selection. Ponder the wealth of anatomical diversity displayed among modern dog breeds: this indicates that at least certain types of morphological changes in development may be rather unconstrained.

Instead, perhaps a more important process motivating stasis, which Steve and several other scientists, including Niles and Bruce, endorsed, relates to the nature of species themselves. Species are broken up into different populations, each often doing its "own thing" evolutionarily (see figure 2.3). The net result of the changes across all these populations may ultimately even out. Think of a one-hit-wonder pop band as analogous to a species, with each of the individual members populations of that species. A lot may happen to the populations (represented by the different band members) of that species, but the species (the band) as a whole doesn't get anywhere and then just disappears. That's a perfect metaphor for the evolutionary history of most of the species preserved in the fossil record.

A CONCEPTUAL HISTORY OF STASIS: THE EARLY YEARS

That life has evolved has of course been completely confirmed by empirical and theoretical results from genetics, systematics—and the fossil record. In fact, not illogically, it was to the fossil record that the early naturalists interested in the history of life turned. They thought that fossils would reveal at least hints of what goes on during the origins of new species, including especially the species we see around us in today's world. This scientific work all started in the earliest years of the nineteenth century. Among the most significant pioneers searching for clues to the dynamics of species origins were Jean-Baptiste Lamarck and Giambattista

Brocchi, both coincidentally named after John the Baptist, who first wrote about evolution and fossils in 1801 and 1814, respectively.

What did they conclude about evolution in the fossil record? First up, Lamarck. He thought that species keep constantly changing. He was studying the fossil mollusks of the Paris basin, now recognized to be Eocene in age and thus roughly 56-34 million years old. (Not knowing the actual ages of his Eocene fossils, Lamarck nonetheless realized they had lived well back in geological time.) Arguing with his colleague Georges Cuvier at the Jardin des Plantes in Paris, Lamarck thought that species did not undergo true extinction but rather slowly and gradually had simply evolved themselves out of existence. Lamarck thought that the species in the modern sea were the modified descendants of those he saw in the Parisian fossil record. However, the in-between states in the gradual transformation of species between the Eocene originals and today's highly modified descendants had simply not been preserved, such that the transition between the ancient and modern species appeared sudden and without gradual transition.

Brocchi thought otherwise. He thought of species as having histories with births (speciation events) and deaths (extinction events), just like individual organisms. Further, he thought species were generally stable during their existence. Aspects of Brocchi's pattern matched up very nicely with modern theory on the evolution of species as seen in the fossil record, i.e., punctuated equilibria. However, we do not postulate true identity between Brocchi's views and punctuated equilibria because Brocchi speculated that the extinction of species is somehow programmed into them, and that they might show some signs of aging near the end of their lifetimes. This notion of species senescence does not jibe with current theory.

He did realize, though, that sometimes species were killed off in the prime of life because of some environmental catastrophe—the geological species-level equivalent of getting hit by a Mack truck.

Brocchi's view of species-level evolution (a.k.a. macroevolution) was also sophisticated. He postulated that species were parts of lineages, which comes close to matching the current modern view of species developed by our good friend Ed Wiley, emeritus professor at the University of Kansas, and referred to as "the evolutionary species concept." To Brocchi younger species replaced their progenitors as geological time marched on, but he stopped short of speculating how this replacement of old species by new, descendant species happened.

STASIS AS DATA

Back to our hypothetical musings of chasing fossils up through time in outcrops, then through larger chunks of time as we visit more quarries in temporally and geographically expanding circles. We are talking here of the rich abundance of fossils left behind by marine invertebrate animals. Sample sizes are often large, meaning that heritable variation can be documented in a single locale (if not always single bedding planes) and compared with the morphology and within-population variation in different places and slices of geological time. Both Bruce and Niles have gone through this process using real-world examples, chasing certain lineages of trilobites (phacopids and asteropyginids) and brachiopods (*Mucrospirifer* and *Athyris*) over considerable spans of time (five-plus million years, centering around ca. 380 million years ago) and stretches of geography (from New York west to Iowa and Canada south to Virginia).

When we did this we ran smack into the stone wall of implacable net evolutionary stasis. Sure, there is always some detectable variation within single samples (if you look hard enough at least!). And yes, even more easily detectable variation when you compare samples from the oldest, middle, and youngest rocks containing these elements of what is referred to as the "Hamilton Group fauna." When all is said and done, though, our fossils looked pretty much the same from their earliest recovered samples right on up through to the very youngest specimens.

Think of it this way: say you go to the beach and collect an assortment of mollusk shells cast up by the tide along the strand line. It's easy to sort them into discrete piles of the "same" shells. And if you go to Cape Cod one day and Long Island on another, you can make pretty much the same assortment of discrete piles of the "same" species. These separate species have pretty much a 1:1 relationship to the populations of still-living species under the waves offshore.

You can do this same sorting game in big-time three dimensions in the fossil record: through prodigious chunks of geological time by collecting vertically, and through space by collecting horizontally over equally prodigious hundreds, sometimes even thousands, of miles-wide chunks of geography, the local beach scene writ large, scaled up strikingly. That's what stasis is: finding those same piles of shells on nearby beaches through nontrivial amounts of time and space.

Back in the 1960s, paleontologists were still being trained to expect to find a pattern, a detectable trace, of slow steady gradual change, at least if their samples were large and spaced closely together. In actuality, with nods to Yogi Berra, they thus couldn't see a lot because they weren't really looking for what they were finding. The expectation of slow, steady gradual change as the hallmark signature of evolution in the fossil record was Darwin's principal legacy to paleontology and the wider circle of evolutionary biologists in general. One of the key points in reference to stasis was that this Darwinian legacy meant that scientists were trained to ignore it, rather than to recognize it as a pattern of special evolutionary significance. The entire history of ignoring stasis as data illustrates how theoretical assumptions can influence the types of data scientists look for and value. This is because scientists are human (at least all the ones we know), and so they behave in very human ways.

Even our good sampling of the fossil record, which is better for modern paleontologists looking in younger rocks, as Lamarck and especially Brocchi had the good fortune or wisdom to work on, failed to reveal the hoped-for slow and steady change that paleontologists were supposed to find.

Ah, the stony evolutionary silence of fossil species that didn't change much throughout their history. The way out was the realization that there *are* changes, marked changes that lead to new species, themselves going on to become stable hallmarks of descendant species, showing up *laterally* (geographically). That's "punctuated equilibria," which we'll hereinafter call "punk eek," the particulars and specific details of its development and discovery told in chapter 1.

The story here is of this key component of net stability displayed by most species from the start to the end of their evolutionary careers. Niles and Steve dubbed this stability "stasis" (Steve, who was very good at coining names for new terms, also came up with the punk eek moniker). Stasis is one of the two main empirical underpinnings of punk eek.

And recovering stasis, instead of the mark of failure by greenhorn paleontologists to find the gradualism that their elders had told them they must find if they did their jobs right, instantly became to us and a few other like-minded paleontologists in the 1970s a phenomenon that cried out for explanation in modern evolutionary biological terms. It took a while for many paleontologists, and even longer for most other evolutionary biologists unfamiliar with fossils to stop insisting that gradualism must be right and stasis itself an impossibility. You really can see a lot just by looking, but only if you're actually looking. As Steve put it, it turns out that "stasis is data." It cannot and should not be ignored, although it often was.

By now, half a century later, stasis has been pretty well recognized, though not always as thoroughly incorporated into evolutionary theory as we think it should be. Furthermore, the incompleteness of the fossil record cannot explain stasis; incompleteness *was* used to explain away the absence of gradual change and the seeming suddenness of change between species sometimes preserved in the fossil record. But an incomplete fossil record just means that there is even more stasis than we're in fact seeing.

STASIS AND TUMULT

Well, if stasis is indeed data, how do we explain it? How in the world can a species remain evolutionarily stable for millions of years, despite some oscillatory wobbling in its features as it plods its way through its time on Earth. (Not really "sleeping," as our title suggests, but sort of "dynamically dormant" for as many as five million years or more.) For it is indeed true, as our friend evolutionary biologist John Thompson points out in his book *Relentless Evolution,*[2] that the genome of any given species is in constant turmoil from place to place and from year to year. Species or their parts are always changing or always moving, yet overall species don't seem to "get anywhere" (just like that one-hit-wonder band) in terms of morphological change, which is, after all, the initial and still very much key problem for evolutionary biologists to explain. Why? That's a key question. Other factors are certainly at play. Maybe even the genome itself, while in constant turmoil, is tossed only within certain limits? Notable scientists like Mark Pagel and Chris Venditti from the University of Reading in England have shown that modifications of the genome do fit the punk eek pattern, with molecular changes concentrated at speciation events, such that molecules mirror morphology. As dynamic journalist Rachel Maddow is fond of saying, "watch this space." Inevitably, more will be revealed as evolutionary biologists and paleontologists continue to explore punk eek.

CHAPTER 3

Survival of the Laziest

Does Evolution Permit Naps?

Yes, to answer the question raised in this chapter's subtitle, evolution does permit naps, if you ask either of us, or if you could have observed the rabbit that used to park itself every afternoon for several successive weeks under Bruce's ornamental maple tree. But we're getting ahead of ourselves. Although we enjoy naps when we can get them, we're also not just about the sleep, and we actually do sometimes stay awake to get inspired by many people, places, and things: Charles Darwin and Miles Davis; fossil grounds in upstate New York and beaches in the Galápagos; Martin Scorsese's *Goodfellas* and George Gaylord Simpson's *Tempo and Mode in Evolution*.[1] We study the pattern of macroevolution to make inferences about the processes that have produced that pattern. We love biodiversity and working in natural history museums. And we love doing scientific research. In this chapter we're going to focus on some research that Bruce was involved with that connects to our long-standing interests in the history of life and the causal factors that influence evolution and extinction. That research further ultimately relates to the topic of what types of activity levels may be most evolutionarily favored in the long term, and it vindicates the argument that humans should conserve resources rather than profligately waste them.

As we mentioned, we are inspired by fossil grounds in upstate New York, and both of us, especially Niles, have spent much time collecting fossils there. But there are many other places to collect fossils.

One especially prolific set of fossil sites can be found along the southeastern coast of the United States. There one can find an exceptional record of marine invertebrate organisms, especially mollusks, the group including clams and snails, from the last 10-15 million years or so. Some of the fossilized shells are so

3.1 Shells of Plio-Pleistocene mollusks from Florida. Courtesy of Jonathan Hendricks, Paleontological Research Institution, and the *Digital Encyclopedia of Ancient Life.*

beautifully preserved that they look as if the mushy, mucousy denizens that once occupied them just died last month (figure 3.1).

We turn our attention to such sites and especially those from the state of Florida and nearby environs containing fossils deposited over the last 3.5 million years.

One of the things that has always intrigued paleontologists and evolutionary biologists and us is understanding why certain species survive and others go extinct. In research led by Luke Strotz, now at Northwest University in Xi'an, China, and Erin Saupe at Oxford University, and including Bruce, we tried to get insight into just this by focusing on how the metabolism or physiology of organisms affects the long-term survival of the species they are contained in.[2] There are all kinds of hifalutin definitions of metabolism (e.g., "the chemical processes that occur within a living organism in order to maintain life") that invoke hardcore physicochemical principles, and if you want to think of it that way, that's fine. We, however, are just as happy if you think of it like your "get up and go," which sadly sometimes may "have got up and left." (Thank you, Steven Tyler of Aerosmith.)

TO SURVIVE, SOMETIMES BEING A SLUGGARD
IS THE BEST STRATEGY

What Strotz and his colleagues found is that the mollusk species that were most likely to survive over the long term contained organisms with statistically lower metabolisms relative to the ones that went extinct. We don't usually think of mollusks as having inherently high metabolisms. After all, a synonym for snail is "sluggard" or "slowpoke," yet some snails and clams are more sluggish and perhaps dilly-dally more than others. In turn, there is reason to believe from ecological studies of modern organisms that those with lower metabolisms tend to live longer than their high metabolism kin. In short, Strotz and his colleagues found a direct link from enhanced survival of organisms to enhanced survival of species. This result shows how microevolution can sometimes translate or extrapolate directly to macroevolution (such extrapolation is not always possible, as we will see). The pattern Strotz and colleagues discovered has been termed "survival of the laziest" in some circles, a phrase that circles back to that legendary phrase "survival of the fittest," which is often thought of as synonymous with Darwinian evolution. It is not. Moreover, it is not a phrase that Darwin invented, although he did subsequently coopt it. Instead, Herbert Spencer, a philosopher who held forth on the evolution of social systems, invented it. It is based on an idea that is now largely discredited in biological and social spheres, since sometimes organisms survive for reasons entirely unrelated to their inherent fitness, such as chance, being in the right place at the right time. In social spheres, historical factors play a far greater role in determining outcomes than evolutionary fitness. Herbert Spencer himself remarked that "in science the most important thing is to modify and change one's ideas as science advances," and that, dear reader, is one of the reasons why biologists don't talk much about Herbert Spencer these days.

Now at this point you might be wondering, "How the heck did Strotz and his colleagues calculate metabolic rate values for mollusks that lived 3.5 million years ago?" It is not as if a long-dead antediluvian snail can be asked to run on a treadmill, but thankfully nothing of the sort is required—and recall that some of the species that Strotz and colleagues analyzed are still alive. It turns out that metabolic rates have been calculated for many groups of still living mollusks by other scientists. If you know the taxonomic group a long-deceased snail or clam specimen belongs to, its body size, and the temperature it experienced, you can calculate its metabolic rate. Just as metabolic rates have been calculated for various mollusk groups, past ocean temperatures (the mollusks being studied were marine organisms) have been determined by climate scientists such as Harry Dowsett and colleagues at the U.S. Geological Survey and Stephen Hunter and colleagues at the University of Leeds and are thus also available. In short, the work performed by Strotz and his colleagues builds on and utilizes earlier studies conducted by previous generations of scientists. Also important in this regard are earlier generations of avocational and professional scientists who collected fossil and modern specimens of mollusks housed today in museums.

HIERARCHIES AND EVOLUTION

Another way of framing this result is as part of a discussion of hierarchies. As a short, apposite digression, there are "things" ("entities" is the more technical term) in biology that all scientists recognize, such as cells, organisms, populations, and species. These exist at different levels of organization, in this case listed from lower to higher levels, and lower levels are nested (or contained) within higher levels. For instance, cells are parts of organisms, organisms are subsumed within species, and so on. Why do we need to think hierarchically to understand evolution? Here we provide only a brief précis as an answer. Notice for now that generations of scientists have asserted the importance of approaching scientific (and other) problems hierarchically, as when the great physicist Richard Feynman remarked: "all the sciences, and not just the sciences but all the efforts of intellectual kinds, are an endeavor to see the connections of the hierarchies."

Each of us has written extensively as to how evolution is the result of interactions between the environment and entities such as organisms, populations, and species. Sometimes evolutionary processes operating at the level of species determine what happens to the organisms contained within them. Other times it is processes acting at the level of organisms that determines the properties of the species they belong to: Strotz and his colleagues found this was the case. Further, physiology seemed to be a paramount process. Metabolism and physiology are about energy use and ultimately energy consumption, especially food and nutrients. Intuitively, having a lower metabolism and using less energy could be seen as a favorable evolutionary strategy, especially during times of environmental change and disruption: at these times less resources and food might be available.

Those organisms that need less resources and less food are more likely to survive during such times, and thus the species they belong to are more likely to survive over the long term.

THERE'S NO EXCUSE FOR BEING A WASTREL

This in no way constitutes an evolutionary justification for human lassitude, or any other justification, for that matter. In fact, a medical doctor from the National Institute of Health (NIH) chastised Bruce after some of the newspaper coverage associated with the Strotz study appeared. The NIH scientist said in as many words, "We've been working for years to try to get people to exercise, and you come along and tell them it's okay to lie around and do nothing." As a paleontologist, Bruce was so astonished that anyone would contact him on any matter of practical relevance whatsoever that it took him a while to respond. He eventually replied to concede that it was regrettable that some had derived this message from the research. Bruce's response continued that a point that the authors had hoped to get across, when it came to humans, was that some of those "lazy humans" consume the most resources, foment the greatest climate change, and favor the ever increasing suburban sprawl and clearing of lands to make way for more domesticated animals and plants, causing the present-day biodiversity crisis. To Bruce's contrite missive there was never a reply. Still, if you're a lassitudinous mollusk or a conservation-minded human your future may be bright, and we thank you for doing your part to conserve resources and help make for a better tomorrow.

ECOLOGY AND EVOLUTION:
SOMETIMES SEPARATE, OFTEN NOT EQUAL

Another result from the Strotz study worth mentioning with respect to hierarchies is the pattern they found when they focused on groups of organisms occurring together in ecological assemblages. As already mentioned, the basal metabolic rate of your average mollusk species alive today is lower than that of the average mollusk living 3.5 million years ago. It might perhaps be reasonable to posit a concomitant decline in the energy usage of ecological assemblages of mollusks over the same time interval. But as with so many things that seem reasonable, things don't always pan out that way. Instead, Strotz and his colleagues found that the total metabolic expenditure of molluscan organisms through time in this region was constant. How can that be, given the demise of more species containing high energy organisms? One possibility is that the total number of organisms increased, even as the number of species fell. If that is the explanation, it is as if there is some fairly constant available stream of resources or energy "out there" in the ecological realm, and one way or another all that energy is going to be consumed. That energy can be totally consumed by many species that each contain few organisms, or it can be by a few species that each contain a plethora of organisms.

A related phenomenon was uncovered by Elisabeth Vrba in 1987. As described in chapter 1, Elisabeth was one of the Three Musketeers of macroevolution as designated by Stephen Jay Gould. She made fundamental contributions to evolutionary biology in general and macroevolution in particular. Elisabeth is a highly incisive thinker. Part of her great success can be ascribed to the fact that she is able to break down any complex scientific problem into its respective component parts. Further, she has a precise, logical mind that always allows her to get to the crux of the problem. She demonstrated these skills in her work on a variety of key macroevolutionary problems that she often viewed through the lens of a group she had a deep understanding of, the Bovidae. Bovids such as cows and water buffalos are a diverse clade of mammals, their total modern diversity some 125 species and their fossil diversity greater than three hundred species. The family also includes antelopes, impalas, bison, sheep, and goats. Most apposite here was aspects of her work comparing the closely related African antelope and impala clades. It happens to be the case that in both the modern biota and the fossil record there were many species of antelope, yet each antelope species has relatively few organisms. By contrast, impalas are of low diversity yet are far more abundant. In short, the total number of organisms in each clade was roughly coeval; they were just broken up into different numbers of species.

What Elisabeth's work shows is that ecological patterns and evolutionary patterns, each fascinating and important, are not necessarily equivalent at the macroevolutionary scale. She was the first to document this at such a scale, though it had been predicted theoretically by various folks, including Elisabeth herself. This result ties back to our earlier discussion on hierarchies. In particular, life seems to be broken up into genealogic and economic (or ecologic) actors. The former comprise genes, cells, organisms, populations, species, and clades; the latter comprise cells, organisms, populations, communities, and regional biotas. Sometimes the twain will meet, as with cells, organisms, and perhaps populations. Sometimes they branch apart. One of the obfuscatory aspects of twentieth-century evolutionary biology was that it often assumed that ecology and evolution were like two lanes of the same highway, running in parallel and leading to the same place. We suspect that's not the case, and Strotz et al. and Vrba each showed that ecology and evolution are more like two roads in a network, ultimately connected but leading to different places. What this means is that evolution and ecology are more complicated than some might have supposed. We should expect that something inherently complex like evolution or ecology can be explained only by an approach that acknowledges and incorporates that complexity. So, we guess, nothing is easy about evolution or ecology. But who said science was always easy? It would have been highly naïve to have thought that it was.

If you don't know Miles Davis's classic jazz album *Kind of Blue*, then you owe it to yourself to give it a listen. It appeared in 1959, at the beginning of a time of major change in this country, its revolutionary and epochal sound fitting to that setting. The implications of any changes to evolutionary theory are by contrast nowhere near as monumental. But the explicit addition of a hierarchical framework *was* an important development for the field. Actually, evolutionary concepts had incorporated a nascent view of hierarchy for some time. For instance, Steve Gould, in his 2002 magnum opus *The Structure of Evolutionary Theory*, argued that Darwin's principle of divergence focused on both organisms and species.[3] In addition, Niles has argued that hierarchy plays a key role in the writings during a critical period of growth of the field, the development of the neo-Darwinian synthesis, when one of its chief proponents, geneticist Theodosius Dobzhansky, argued that evolution wasn't "just mutation" but also involved phenomena particular to populations of organisms. Elisabeth focused on the extent to which the number of species in a group increased through time not due to the inherent fitness of their component organisms but rather due to their inherent propensity to speciate.

The more explicit incorporation of hierarchy into evolutionary biology was directed in an important way by the activities of us Musketeers. Steve was obviously a big part of that, and he definitely advocated for a hierarchical view that considered species to be fundamental units of evolution. The Three Musketeers were no reductionists. However, unlike Niles and Elisabeth, he was significantly less salubrious about the existence of ecological hierarchies or even the distinction between ecology and evolution. This may have been partly an outgrowth of the fact that Steve was just not very interested in or may not have been convinced of the broader significance of the field of ecology, especially as it was applied to the study of the fossil record, a subdiscipline referred to as paleoecology. He said as much to several of his former graduate students who are now highly active faculty, including Warren Allmon and Linda Ivany, and he made similar statements to Bruce as well.

Although we diverge from Steve on this point, we can perhaps see where he was coming from. In the 1960s and 1970s paleoecology was a subdiscipline adrift from the rest of paleontology. Most of its practitioners were simply looking to find the same patterns in the fossil record that ecologists were finding in the modern. They were also trying to use the same theoretical perspective. However, this created a significant disconnect, as paleoecologists should have recognized that the timescales they were considering were many orders of magnitude greater than those ecologists studied. Thus they should have expected to find different patterns and employ disparate theories. Paleoecology did not have to be handmaiden to ecology and instead could explore its own distinct intellectual and theoretical space. Until that was recognized by paleoecological practitioners, the discipline would never live up to its scientific potential.

Crucially, paleoecologists started to do just that in the 1980s. One of the early important figures in this area was William Miller, who among other things argued that what paleoecologists were calling succession was almost always very different from what ecologists referred to: in ecological parlance succession denoted the changes in vegetation that could be observed in an empty field over the course of decades. These types of changes will rarely if ever be observed in the fossil record. However, changes among fossil assemblages can be witnessed that take place over the course of hundreds of thousands of years; these were not succession and instead reflected long-term responses to climate change. And understanding of these is of course highly relevant for understanding what will happen to our planetary biota in the future because of human-induced climate change.

Paleoecology has become a much more vibrant area than it was when Steve was in graduate school. A scientist always finds it a challenge to keep up with new developments. It's something that every scientist struggles with. It's not about being lazy, or survival of the laziest. Instead, it's about maintaining one's sanity—more and more new science comes out all the time, and the pressure is always on for a scientist to publish new science. That can leave little time to read about or absorb new developments. This is especially true for a scientist who was as busy and hard-charging as Steve was, consistently publishing numerous books and giving talks all around the world. Forget about lazy: we know of no one who worked as hard.

Notably, one of the more significant advances in paleoecology, the development of the concept of "coordinated stasis," which argued that ecological and evolutionary turnover in the groups of species found together in the fossil record occurs relatively rapidly and following long periods of stability, was led by some of Steve's erstwhile graduate students: Linda Ivany, Ken Schopf, and Paul Morris. This work was done in collaboration with Carl Brett, who was an expert on the fossil record of the Devonian of New York State, which provided the type example for coordinated stasis. Steve also had other students who made important contributions to paleoecology, among them Warren Allmon. Even though Steve might not have always appreciated ecology, he was smart enough to recognize that his students were onto something. Sometimes the failings of a parent do not adhere to their progeny. And even among what others may see as a closely aligned group of scientists, there can be substantial differences of opinion. After all, ecology is a pretty big thing! Just to think that even the members of the Three Musketeers might not always be on the same page gives you an idea of what a disparate intellectual space practitioners in the fields of paleontology and evolutionary biology can occupy.

CHAPTER 4

Time's Arrow, Time's Cycle, Times Square

Gould, Kant, and Super Dave

Bruce's introduction to quasi-stuntman Super Dave Osborne was in a narrow, dimly lit Hell's Kitchen dive bar in the late 1980s. All New Yorkers know that Hell's Kitchen is situated not far from the glittering lights of Times Square, and Times Square was a very different place back then. So was Hell's Kitchen, for that matter.

Science has also changed since those days, although science was not Bruce's main focus at the time: he had not even applied to graduate school, though soon he would do so, with the aim of working with Niles, who indeed was well ensconced in his career. From the depth of grime on the bar's wooden fixtures and floor as he recalls it, Bruce estimates that the bar had likely stood there since the 1940s. That's where Bruce had his first and last encounter with Super Dave. Last, that is, apart from watching some *YouTube* videos viewed while researching this essay, which we recommend that you do, especially the mad stunts he performed on his series *The Super Dave Osborne Show.*

Now, for those that don't know him, Super Dave—his real name is Bob Einstein, brother to Albert Einstein, who goes by the name of Albert Brooks—comfortably occupied the intellectual space between a dilettante and a fool. However, in that intellectual space, the legacy of Super Dave leaves us lessons for the process and practice of science.

Bar televisions typically broadcast bland or depressing visual fare, in the absence of a sporting event, and the TV in this bar seemed to be no exception. As Bruce looked up from his bottle of Budweiser, he saw a middle-aged, moderately athletic man on the tube, clad in an Evil Knievel-style jumpsuit. Super Dave sat astride a medium-horsepower motorcycle, which bore the cheeky moniker "The Super Dave Atomic Bike." During the broadcast, another patron roughly Bruce's age also sat at the bar; for all Bruce knows, today that patron might be a major motion picture executive. We don't know if shows were better back than or not; we wonder what the possible motion picture executive would say? What Bruce does know is that he was not especially riveted to the telly, but he was riveted enough to notice that Super Dave held a helmet in his lap, while a small shack engulfed in flames sat nearby.

Super Dave then looked into the camera while a straight man clad in a sport coat said something to the effect of, "Now, for his next stunt, Super Dave will drive through this burning building." Super Dave then put on his helmet, revved his engine, rolled around to the back end of the shack, and crashed through. A scant two seconds later he burst through the front door, still on his motorcycle. This may not have been the stupidest thing Bruce had ever seen, but it was certainly up there. The other bar patron said, in language cleaned up for the purposes of our PG-rated book, "this guy is the biggest idiot on the planet!"

But then something happened that changed everything. Super Dave took off his helmet; the camera zoomed to his face. There was Super Dave, flashing a toothsome smile, but he was also now sporting a pair of large, false eyelashes. He whimsically batted these at the camera. The two of us shouted something to the effect of, "This man is brilliant: Super Dave rules!"

How can someone go from doofus to hero or vice versa in a matter of seconds? Where is the line between disdain and respect, how are they earned, and why can that line be so easily crossed? It's worth remembering Nigel Tufnel, a key character in the 1984 movie *This Is Spinal Tap*, who retorted, "There's a fine line between stupid and clever." This brilliant saying, we think, belongs on the shelf next to Ernest Shackleton's glacially cool comment that "there is a fine line between success and failure" as he explained how he and his crew survived one of the most harrowing Antarctic expeditions ever undertaken.

THE IMPORTANCE OF BEING ERNEST, SHACKLETON

Sir Ernest Shackleton, for those of you unfamiliar with him, was a famous polar explorer from the tail end of the halcyon days of polar exploration. He and his crew left for Antarctica aboard the ship *Endurance* before the start of World War I. The ship and its crew got stuck in Antarctic pack ice, and their only recourse was to ultimately leave the ship, first to float on pack ice for many months and then to take three small boats in search of islands along the coast of the continent. Hundreds of miles floating, rowing, and hunkering in the trackless ice and oceans around Antarctica in horrendous weather conditions was far more hazardous than a jaunt through Times Square ever was, even when the area was at its very nadir. Shackleton and crew eventually reached Elephant Island, but their saga was not over. There was little to eat, the climate was inhospitable, and the island was devoid of potential rescuers. Their only hope was to head farther north by boat to the site of nearest human occupation. So Shackleton and a quarter of his crew took the skiff *James Caird*, which was only some twenty feet long, north from Elephant Island into the teeth of the wind-tossed Southern Ocean and its crashing, gigantic waves.

They had limited supplies, and their only means of navigation depended on being able to accurately sight the sun, which, because of the grim weather, exhibited only feeble luminosity at best. Their destination was a small colony of whaling stations on South Georgia Island, a tiny speck of rock more than seven hundred miles away. Today it still seems extremely remote, and it is the only principality on Earth that celebrates "Toothfish Day." This holiday honors the Toothfish, sometimes called the Chilean sea bass, which is now sadly endangered due to overfishing. In 1916, South Georgia would have seemed even more remote, and if they were to miss it they would be adrift in the South Atlantic with no prospect of salvation.

Most of us in our day-to-day lives cannot fathom, and thankfully do not have to fathom, how our entire existence is balanced on the edge of a knife. This is an aspect of the modern world, where we, at least in the so-called developed nations, can largely ignore many aspects of natural selection for most of our existence, even during a pernicious pandemic, until we become very old. Throughout our lives we walk on such a knife edge, and at different times or under different conditions the width of that edge varies. Eventually we've got to fall off, for death is inevitable. As paleontologists well know, death is often though not always precipitated by natural selection, a statistical law—meaning that death simply has to happen, and that can be pretty disastrous, especially if you become a statistic.

What if your life came down to such a knife edge? How would you react? The song "The Impression That I Get" by the Mighty Mighty Bosstones provides thoughtful musings on this very topic, so if you have the chance, check it out.

Sadly, for some the edge can be exposed earlier in life. One's reaction is said to illustrate character, but what does that mean anyway? When reaching such a near singularity, an apt response might be: "It's hopeless." That was dimly remembered to be the epithet frequently intoned by the character "Glum" on the cartoon series *Gulliver's Travels* that Bruce watched as a child.

Yet that wasn't Shackleton's reaction. He always remained positive and supportive. He presented a model of cool and aplomb that calmed and motivated his crew and maximized their chances of survival. Ultimately, after their skiff reached South Georgia, without equipment and in somewhat ragged clothing, they still needed to climb and then descend a sizable mountain peak replete with glaciers and icefields to reach the nearest South Georgia whaling camp. When they strode into that camp, they realized they had made it, but would soon learn that millions of others teetered on the brink amid the conflagration of World War I.

STEPHEN JAY GOULD AND LIVING LIFE ON THE EDGE

Aspects of the characteristic confidence and commitment to people and preparation that Shackleton exuded, what maximized his chances and those of his crew, also adhered to one of the most misunderstood recent figures in the history of science, Stephen Jay Gould. To Niles, Steve was a fellow graduate student, a friend, and a noted collaborator; to Bruce he was a mentor and teacher. We respect the dude a lot, but that doesn't mean that we agreed with him on each and every last thing. Friends, mentors, and mentees can have differences of opinion, as long as they show respect, and we definitely aim to show it here.

When it came to awesome life challenges, Steve experienced one in the early 1980s when he was diagnosed with mesothelioma, an extremely virulent and often fatal type of cancer. Can we truly compare that to Shackleton's voyage? We believe we can, by analogy, just as Ernest Hemingway used the voyage of a single individual struggling aboard a boat in *The Old Man and the Sea* to illustrate how we are all so perilously alone.

It's good not to be lonely, because loneliness reminds you that you are alone, but lonely or not, we are all truly alone. Moreover, we all stand somewhere on that metaphorical knife edge. Most times we don't know precisely where we sit on the blade or how narrow the edge is, and that ignorance is generally blissful. Consider the 500-million-year-old trilobite (figure 4.1) whose mortal remains are left behind in a layer of sedimentary rock.

4.1 Image of the ~500-million-year-old trilobite *Elrathia kingii*, an ancient denizen of Utah, in the collections of the University of Kansas, Biodiversity Institute, Division of Invertebrate Paleontology (KUMIP 521018). Photo by Natalia López Carranza (KU).

It was unlikely to have been troubled by the fact that the cocoon of sediments surrounding it represented its burial shroud, and that must have made it metaphorically easier to be trilobite than human.

When Steve found out he had come to the very thinnest part of the knife edge, he showed a dogged determination that might have made Shackleton proud. In particular, Steve showed an unwillingness to hang up his spurs on the scientific research that mattered most to him, and that preparation, that coolness, that unwillingness to let place and time matter is what makes one a clutch player. That's the one that you want with the ball in their hands when the game clock is winding down and you need them to take the tying or winning shot. Gould wasn't a leader of people, at least the way Shackleton was; his charisma came from his prodigious intellectual output and his rhetorical skills. But he displayed personal leadership genius not only during the "normal" stretches of his career in academia but also during times of extreme difficulty. Over the period from mesothelioma diagnosis to recovery Steve continued to publish, and ultimately produced a tremendous tide of high-quality, well-cited research, some that lapped on to the subsequent years of his career. Let's focus on one by-product of work from that stage of his career, the one that conjoined Steve to Super Dave.

Stephen Jay Gould certainly differed from Super Dave in several key respects, but there were also similarities. Like Super Dave, Steve had a whiff of the dilettante to him, though Steve notably straddled the line between it and genius. Unlike Super Dave he was entirely unathletic. However, Steve also knew the value of cerebral legerdemain and how a seemingly simple trick could provide force to an otherwise standard academic exercise. Nowhere was this better illustrated in his career then during his lengthy flirtation with the topic of time's arrow and time's cycle, which eventually culminated in his eponymous book that came about not long after Steve faced and stared down his experience with mesothelioma.

TIME, SCIENCE, AND STAMP COLLECTING

Steve dignified the metaphors of "time's arrow" and "time's cycle" as critical juxtaposed and distinct ways of thinking about the history of the earth and of life. Over the years he first referred to Time's Arrow as an "idiographic approach," and he later discarded that in favor of a "contingent approach." Steve sometimes substituted "nomothetic approach" for time's cycle.

Contingency, idiography, time's arrow: these are all just fancy terms for history, that is to say, the recognition that things go from a beginning to an end, from first to last, from alpha to omega. That's what those terms meant to Steve, and that's generally what they mean to other folks as well. Why is the fact that there should be contingency or history in the history of life a big deal to anyone? A vast swath of the scientific community posits that historical phenomena and processes are a troublesome type of noise that interferes with and is outside the direct purview of science. This argument that "history bad, let's ignore it, experimentation good, let's do it" is presented by those who frown on the meaning or significance of unique events and natural history studies for deepening scientific understanding. These are the same type of scientists, sometimes joined by philosophers, who refer to paleontology and other types of natural history studies as stamp collecting, not only casting aspersions on paleontology but also on the hobby of philately.

Why is there such a dramatic bias against stamp collecting, while other hobbies like coin collecting get a pass? The answer is unclear; perhaps numismatics is sustained by a stronger lobbying organization. But it was even a view that Steve himself supported in 1977, "the task of history is to explain the contexts so clearly that they can be separated and subtracted, thus permitting us to see the unchanging themes."[1] In short, he started out "loving the Time's Cycle, the Time's Arrow not so much." Later he changed his mind, in fact, several times, his opinion oscillating back and forth like a clock's pendulum.

Beyond the rather dastardly view of philately in certain scientific and philosophic circles, it appears that some scientists and philosophers actually believed that if an area of study and knowledge involves having to consider historical events, then it is an area of study that lies outside the realm of science. They claimed this because they posited that, absent repeatability and the potential for experimental manipulation, one cannot say why something happened, only that it did.

By contrast, nomothetic approaches, which emphasize the search for eternal laws, what Steve referred to as time's cycle approaches, were elevated by these scientists and philosophers. In a general sense they seemed to think that because an atom is posited to have behaved the same today as it did back in the Devonian and seems to behave the same whether it's on the moon or in the middle of the ocean, it meant that the study of atoms is in some way better, more important, or more dignified than the study of trilobites; trilobites, of course, behaved differently in the Devonian—they were alive and thus actually behaving, for one thing—and would be far from their acme of existence on the moon. Further, trilobites are not amenable to experimentation, having been dead for 250-plus million years. Therefore, these scientists and philosophers claimed, no laws or deep scientific principles could emerge from the study of trilobites, or indeed of any natural history phenomena.

Regardless of the acerbic grumpiness toward philately, natural history, and paleontology in certain scientific and philosophic quarters, Steve really cared a lot about time's arrow and time's cycle. He categorized them as intransigent dualities or unique yins and yangs in the quest for knowledge. We agree that intransigent dualities can exist in science: for instance, you can't measure the position and direction of motion of a quantum particle at the same time, the so-called Heisenberg principle of uncertainty. Is the duality Steve dignified similar? To cut to the chase, we think not. We think that Steve might just have been confused or paying too much attention to those grumpy scientists and hifalutin philosophers, who were themselves confused. We think that Heisenberg was confused about a lot of stuff too: he did, after all, at least partially cozy up to Hitler's Nazi regime. But Steve's confusion proves instructive.

THE MBL GROUP

When did Steve start his romance with time's arrow and time's cycle? Although it was a topic he considered very early in his career, it truly began in earnest with his participation in the Marine Biological Laboratory (MBL) group, a collaboration including the distinguished scientists Tom Schopf, David Raup, Jack Sepkoski, and Daniel Simberloff. Each member of the group demonstrated their genius in numerous ways, including the fact that they managed to figure out a way to use the discussion and analysis of paleontological data as an excuse to get to hang out all expenses paid at the MBL over the course of several summers—and it doesn't hurt that the MBL is located in beautiful Falmouth, Massachusetts, along the southwestern edge of Cape Cod.

David Sepkoski wrote the 2012 book *Rereading the Fossil Record*, which provides an excellent, detailed discussion of Steve's (and the other scientists') participation in the MBL group, which is quite relevant to this topic.[2] We highly recommend that you read it.

One of the reasons the work of the MBL group was significant is that it employed the power of simulations to endeavor to place paleontology into the sphere of experimental science. One of the key activities of the MBL group was to use computers to take idealized, artificial groups of hypothetical species and cause them to diversify and go extinct through simulated millions of years of evolution in just a few seconds. (These results were published in several papers.)[3] In scientific parlance, the MBL group was generating numerous time series using an evolutionary model. They used constant probabilities of speciation and extinction in their simulations to produce numerous iterations of a largely randomized evolutionary process to see how hypothetical groups of species might evolve. Then they were comparing those numerous, hypothetical iterations to actual patterns seen in real fossil groups, like trilobites. Significantly, they found that several of their simulated clades showed diversity patterns congruent with those found in the fossil record. For instance, some showed dramatic early increases in diversity with later stabilization. Others showed dramatic declines that looked like major extinction events. In short, by simulating repeated iterations of evolution they were trying to insert an element of experimental repeatability into the analysis of paleontological time series. The MBL group had thus gone toward inserting a nomothetic element into paleontology. Indeed, there is a well-known 1980 paper by Stephen Jay Gould that recapped some of the discoveries of the MBL group and advocated for more focus in paleontology on nomothetic approaches.[4] Notably, there he derided the fact that paleontology as a discipline seemed more focused on identifying "irreducible historical uniqueness" and was too much about finding contingency. In short, in 1980 he appears all in on the nomothetic side. It is noteworthy that by the late 1980s this was a view he would completely renounce.

KANT STAND YA

At this point it is worth mentioning the debate about the difference between nomothetic and contingent approaches in philosophical discourse. To do so, we need to turn to two figures, Wilhelm Windelband and Immanuel Kant.

In 1894, during what is known as his "Rectorial Address" at the University of Strasbourg, Wilhelm lectured on how scientists sometimes focus on properties of individual things (i.e., contingent approaches) while sometimes they focus on general principles (i.e., nomothetic approaches).[5] We find it completely fascinating and indeed downright astonishing that anyone showed up to a lecture in Strasbourg (or anywhere) with that title, and we are absolutely blown away by the fact that the contents of this lecture were written down.

One critical reason why Windelband got good subsequent press on this address/lecture was that he was trying to reevaluate certain topics that had been treated in Kant's *Critique of Pure Reason*. And that is truly evocative, because we are first of all amazed by the genuine gall of Kant to have simply woken up and one day decided to start writing an eight-hundred-odd page book (ultimately published in 1781) *critiquing pure reason*! If someone is going to spend eight hundred pages critiquing any type of reason, then by definition they are doing something entirely *unreasonable*. However, this is a matter that truly pales in significance to the fact that if Windelband had simply rebranded his lecture "Critique of *Critique of Pure Reason*" rather than "Rectorial Address" it would have been far more famous today. This is why, if we wanted to be even more pedantic, we should have subtitled this chapter "A Critique of a Critique of a *Critique of Pure Reason*." Take that, Wilhelm and Immanuel!

Immanuel Kant's *Critique of Pure Reason* is, among many things, a diatribe against treating historical subjects, including the history of life, in a scientific fashion. He says its acquisition and analysis is "not scientific and thus inconsequential," which to Kant is one of the worst things you can say about something, and coincidentally is the same excuse we've used to avoid watching the Grammy Awards since 2000 or so. Immanuel thus would have devalued paleontology, a discipline that is assuredly the type example of a historical science, and viewed it as scientifically lacking in critical ways. That is why we and all other paleontologists owe a debt of gratitude to Windelband, for in his 1894 address he rectored away about why historical sciences, including natural history, have value.

We should note that we have simplified things drastically here. For instance, in 1988 a psychologist named Albert Silverstein elucidated how the debate about historical sciences has a rich and extensive history among philosophers going back to the time of Aristotle.[6] This brings to mind Bruce's New York City public school social studies teacher in junior high school back in 1976, Mr. Silverstein, who had two bits of sage advice: buy shares of Standard Oil stock as soon as your parents would let you, since they were sure to go up (in price), and don't throw wooden chairs out of fifth floor school windows, because they were sure to go down (onto the streets of Chelsea below). Bruce followed only one of those bits of advice, and it is for this reason that he is neither wealthy nor possessing of a criminal record, unlike other children in the class.

Steve probably couldn't stand Kant, at least on the issue of historical sciences. Still, Steve would never have behaved like the gym teacher in the classic television show *Seinfeld* who cruelly tormented George Costanza by chanting "Can't Stand Ya." Yet in 1968 he remarked that "it is often said that natural history is science of a lower order than experimental biology. If description and explanation are viewed as ascending orders of sophistication, this charge is valid when natural history remains at the level of 'plain story,' for the very undertaking of an experiment implies a search for explanation. Yet there are a variety of natural situations that possess the essential character of experiments, even though no human manipulation of material is involved."[7] Further, any evolutionary biologist or paleontologist worth their salt would agree that there are timeless evolutionary laws. Consider natural selection or allopatric speciation, both notably fully appreciated only long after Kant's 1781 book.

Returning to the MBL group, one of its important premises was that to make any statements about why any group seen in the fossil record had so many or so few species, first those patterns had to be compared with patterns that could be produced at random by what is called a random walk, previously mentioned in chapter 2. During a random walk, a thing drifts aimlessly and without direction through space. This sounds like a pleasant way to spend many a sunny afternoon. Building on that, a random walk of the number of species in a group would treat the chance that a group's diversity increased, decreased, or stayed the same as equal. Even if it follows a random walk, the diversity of a group that starts out as a single species can reach sizable values, especially if early in its history it attains a value above a certain threshold. By contrast, a group that stays at a low diversity level is always at risk of going entirely extinct.

By focusing on random walks, the MBL group was not necessarily always arguing that evolution was random, although they did suggest that random factors played an important role in the history of life. They were generally suggesting, however, that a complex constellation of factors might produce a given evolutionary pattern such that these factors could meaningfully be studied as aspects of a random stochastic process. In their scientific papers the members of the MBL group argued that one of the significant impediments to the success of paleontology as a scientific discipline was its focus on why any individual taxon went extinct at any given time. Thus, they were suggesting that the contingency-based approach to paleontology was antithetical to its scientific status. However, David Sepkoski, in his 2012 book, nicely described how there was not unanimity of opinion among the different participants in the MBL group. Tom Schopf especially favored the nomothetic view on paleontology, whereas Steve sometimes valued contingency.

Some of the major conclusions of the MBL group were that differences in the diversity history of actual fossil groups need not imply any innate biological differences between them and that evolutionary histories of simulated groups operating under the same conditions can differ dramatically. This is significant because it basically reveals the importance of taking contingency into account when using the nomothetic approach to study the history of life. Simply by historical accident (or contingency), groups can show dramatic differences in their evolutionary patterns over time, even if there are no differences in the evolutionary processes generating those patterns. In essence, this is one of the ways that contingency factored importantly into the history of life: complex systems often behave chaotically, with simple differences in initiating conditions leading to dramatic differences over the long term. Steve later explained this using his metaphor of the tape of life: "if you take life's actual tape and run it backward, would you know that the story could not happen this way."[8] By the mid- to late 1980s Steve was breaking up with nomothetic approaches and time's cycle and beginning a romance with the time's arrow perspective.

Early on, Niles was also very much involved in this debate as well. For instance, Niles published a reply and commentary to the works and perspective of the MBL group all the way back in 1976,[9] while Bruce was enjoying Mr. Silverstein's class and learning to navigate the distinctive social milieu that was the New York City public school system of the 1970s. We are of course biased, but we believe that the perspective Niles espoused in his 1976 reply and that he subsequently revisited and amplified in his 1992 book *Interactions: The Biological Context of Social Systems* with Marjorie Grene and his 1999 book *The Pattern of Evolution* provided the solution to the problem that so vexed Steve. We will come to that in due course.

TIME'S ARROW, TIME'S CYCLE: THE BOOK

Now let us turn to Steve's eponymous book on the topic. It truly is an excellent book, and we recommend that you read it. What makes it exceptional is that it draws in all kinds of neat stories and metaphors, and it makes the history of geology, as arcane a subject as any, seem interesting and relevant to modern day readers. Also important in any day and age, the reviews of the book were stellar. No less a personage than John Updike, in perhaps the most august of all literary institutions, the *New Yorker*, praised "Gould's lucid animated style" that displayed "a passion that approaches the lyrical." For any author, that's just about the zenith of existence. To use a metaphor from Steve's great love of baseball, he certainly hit it out of the park.

Even though he called the book *Time's Arrow, Time's Cycle*, seemingly emphasizing polar opposites, in the book he went so far as to recognize that "all dichotomies are simplifications, they are not true or false, simply useful or misleading" and "time's arrow and time's cycle both capture important aspects of reality."[10] Some of the choicest parts of this book involve Steve's efforts to reconstruct the early history of debate among geologists on this topic. For instance, he dignified James Hutton and Charles Lyell, the latter a subject of chapter 11, as two of the important geological exponents of a time's cycle perspective, partly for their work documenting that Earth was incredibly old. Yet he also criticized Hutton, who took an even more ahistorical view than Lyell, for not grasping "the power, worth, and distinction of history."[11] In essence, Hutton had strived to make geology like Newtonian physics, but it could and should never be like Newtonian physics, a view advocated by Niles in his 1992 book with Marjorie Grene. (In that book Niles and Grene used the term "mechanical philosophy" when they referred to the ahistorical view.)

Steve began his 1987 book with figure 4.2, the illustration from the frontispiece of the 1664 book *Telluris Theoria Sacra* or the *Sacred Theory of the Earth* by Bishop Thomas Burnet. Steve showed there his skill as a showman and indeed channeled the "ostentatious sensibilities" of a Super Dave. An evolutionary biologist and paleontologist that reprints an illustration showing Jesus standing astride worlds, from a 1664 book written by someone who had an obviously theistic perspective on the history of the earth, is doing something akin to flashing false eyelashes after zooming through a flaming house and crashing out the front

4.2 Frontispiece from *Telluris Theoria Sacra*, from the Wellcome Library, London, http://creativecommons.org/licenses/by/4.0/.

door. It's never done for any reason other than showmanship, like the lighted billboards and flashy tableaux of Times Square. It may well represent the mid-1980s equivalent of virtue signaling. It also inspires awe, for tracking down a book published in the seventeenth century is never easy, nor is relating one to modern scientific discourse. In any event, our hats are off to Steve and to Super Dave. They did something tricky, but we're just not sure that they really needed to do it.

Steve's views here had seriously evolved from those circa 1980. In 1980 he never would have argued that history was important just for history's sake, but in 1987 he did. However, in 1987 he hadn't entirely abandoned time's cycle in favor of time's arrow just yet. Steve recognized that "uniqueness is the essence of history, but we also crave some underlying generality, some principles of order transcending the distinction of moments."[12] Both time's arrow and time's cycle matter. In certain respects this can be regarded as one of Steve's attempts at developing a weltanschauung, a worldview. Explaining evolution, in such a view, was about abstracting statistical generalities (or processes) from life's pattern, and then extracting the historical factors that perturbed life from a random or stochastic trajectory.

WONDERFUL LIFE AND BEYOND: THE JOURNEY CONTINUES

Steve's views on the relative importance of Time's Arrow and contingency for our understanding of evolution reached their high-water mark in his 1989 book *Wonderful Life* and other publications in the mid-1990s (however, yet again they would recede from this apogee, with a return to a view that countenanced a greater role for time's cycle, as we shall see). Consider his statement that we "should treasure the intricate story of our planet and its life." He even asserted in a follow-up to this book that "the discovery of timeless and universal laws and the prediction of all occurrences under their guidance cannot be an expectation or even a desideratum."[13] Was Steve suggesting that there were no general unifying principles or mechanisms governing the history of life? Certainly *in toto* Steve did not believe this, as he continued to argue in various venues for the importance of punctuated equilibria (and thus also allopatric speciation). But in the late 1980s up through the middle 1990s he seemingly had partly given up on the search for unifying principles and instead came to emphasize individualistic particulars.

This was codified in Steve's 1989 argument in the book *Wonderful Life* that if we could rerun the history of life—replay the tape of life, in his parlance—we would obtain a very different result. Steve seemingly felt the need to act as a cheerleader for paleontology to other biologists, and to do so he felt he needed to emphasize the paramount importance of history, "historical science is not worse, more restricted, or less capable of achieving firm conclusions because experiment, prediction, and subsumption under invariant laws of nature do not represent its usual working methods. The sciences of history use a different mode of explanation, rooted in the comparative and observational richness of our data."[14] He admitted that he used to think "all science demanded timeless statements based on universal laws," but was now convinced that "pure narrative, well done, ranks among the highest forms of science."[15] Here he was challenging a more general societal opinion well captured by conjoining the casual musings of two incisive minds: "history is a set of lies agreed upon" (Napoleon), and "there's math, and then everything else is debatable" (Chris Rock).

The view of these luminaries was that history is too open to interpretation, while math always lay farther up the fairway of intellectual merit and prestige. Of course, the critique of historical sciences by certain scientists and philosophers pays no never-mind to the fact that a paleontologist such as Steve might be just as or even more likely than a chemical biologist to use mathematics, but it seemed to be built chiefly on the notion that somehow trilobites were less rigorous as data points than titanium atoms.

In case you were thinking that Steve might hold onto this view on the topic, one needed only wait until six years after *Wonderful Life* was published to witness another swing of the pendulum. In particular, Gould returned to the view propounded in *Time's Arrow, Time's Cycle* and suggested that the "events of a complex natural world were divided into two broad realms—repeatable and predictable incidents of sufficient generality to be explained as consequences of natural law and uniquely contingent events that occur in a world full of both contingency and genuine ontological randomness as well...[and] contingent events mistrusted and downgraded by traditional sciences should be embraced as equally meaningful." Much of the history of life might be unpredictable, but this is due to the natural character of the world. Gould went on to add that "nature's laws and history's contingency must work as equal partners in our quest to answer 'what is life?'"[16] He expressed a generally similar viewpoint in his 2002 magnum opus *The Structure of Evolutionary Theory*, though at the very end of the book he seems to be back to where he was at in his 1980 nomothetic approach paper: "I always thrilled more to the power of coordination than to the delight of a strange moment."[17] We also should note that we sympathize with key aspects of this quote as we enjoy being as coordinated as we possibly can, and we try to avoid strange moments.

MOVING BEYOND TIME'S ARROW AND TIME'S CYCLE

Steve started out more sympathetic to time's cycle, later endorsed a view that accommodated both time's arrow and time's cycle, subsequently moved all in on time's arrow, later returned to a view that valued both, and finally retorted in the last pages of his great 2002 book that time's cycle is better after all. The dynamic is a bit dizzying, maybe even cyclical. But folks are absolutely allowed to change their mind, and certainly Steve spilled a lot of ink on this subject. His intellectual trajectory on this topic reminds us in some ways of the very same random walks that the MBL group singled out for study. We note that many years ago evolutionary biologist John Maynard Smith said something very similar about Steve to Niles, although Maynard Smith was not necessarily Steve's biggest fan.

Now that we've considered Steve's perspective, we'll move on to what some others have thought about this issue as well. Niles is on the record as saying that the entire contingent versus nomothetic debate, what Niles referred to in his 1989 book *Macroevolutionary Dynamics* as the distinction between historical and functional sciences, was neither valid nor productive. Indeed, building on that work and Niles's 1999 book *The Pattern of Evolution*, the perspective we are adopting here is that no matter what type of science someone is doing, whether physics or paleontology, biochemistry or botany, you're still observing patterns and using those observations to make inferences about processes. That is, the only way to derive nomothetic principles is to study contingent realities such that time's arrow and time's cycle cannot be separated, since history is a natural experiment. When paleontologists study large-scale evolutionary patterns preserved in the fossil record, they don't need to settle with "just" chronicling evolution; they can also discover causal processes. From the perspective of macroevolution, the big picture study of evolution, repeated analysis of contingent histories is the key to discovering whatever nomothetic principles exist in the history of life.

Indeed, one key research area in evolutionary biology involves reconstructing the evolutionary history of groups of species to make inferences about the processes that most influenced that history. Both of us have worked in this area, termed phylogenetics, for a long time, and we know enough about it to say that it developed via the contributions of many scientists, including Willi Hennig in his 1966 book *Phylogenetic Systematics* and our good friend Ed Wiley in his 1981 book *Phylogenetics*. Bruce had the privilege of being Ed's coauthor on the 2011 second edition. In our view, as we say there, "historical groups function significantly in science because they are the result of the operation of natural processes on their parts" (113). All patterns are imbued with one or more processes, and thus the contingent vs. nomothetic distinction crumbles.

HUME IS WHERE THE HEART IS

This perspective shares much in common philosophically with David Hume's assertion in part I of *Liberty and Necessity*, in *An Enquiry Concerning Human Understanding* (1748) that "history informs us of nothing new or strange in this particular. Its chief use is only to discover the constant and universal principles of human nature." This also partly diverges from some of Hume's other views, because he weighed in extensively on the topic of inferring causation when numerous times one thing happens and every time it does a second thing results. Does such a circumstance mean that the first event causes the second? It might, but one can never prove that. This is one of Hume's great insights.

We are not alone in admiring Hume. No less a brilliant philosopher than Bertrand Russell, in his epochal 1945 book *A History of Western Philosophy*, remarked that "in his direction, it is impossible to go further. To refute him has been, ever since he wrote, a favourite pastime among metaphysicians. For my part, I find none of their refutations convincing."[18]

Turning to one of the key areas where Hume truly shone is apposite to our theme here, since he made conclusions about the scientific pastime of using data from the past to make inferences about the future in areas beyond pure math. Hume was quite intrigued by the question, "How do we determine cause and effect?" To Hume only math provided certain knowledge, as Chris Rock has told us.

To gain some insight as to why Steve was so troubled by the topic of time's arrow vs. time's cycle whereas we are not, we will delve into Hume's philosophy on causation to see how he got there. In essence, Hume did not even buy thinking about "cause" because "the inference is not determined by reason, since that would require us to assume the uniformity of nature, which itself is not necessary, but only inferred from experience. . . . The frequent conjunction of A and B is no *reason* for expecting them to be conjoined in the future, but is merely a *cause* of this expectation."[19] What this means is that Hume should be cited as the source for all the disclaimers that appear at the bottom of every investment prospectus and website: "Please remember that past performance may not be indicative of future results. Different types of investments involve varying degrees of risk."

Further, the same perspective on *reason* holds true for experimental biology: even though an experiment produced a particular result one time or a thousand times, it still could produce a different result the next time it was performed.

Don't get us wrong, we're not knocking experimental biology. We don't think *that* experiment is going to produce a different result, whatever *that* result may be. Indeed, we would say that if enough data are gathered, or enough experiments are performed, then you've got to like your chances that the next resultant pattern will be as predicted, but this, as Russell so eloquently stated, is based on "the principle that those instances, of which we have had no experience, resemble those of which we have had experience" and "this principle is not logically necessary, since we can at least conceive a change in the course of nature."[20] Instead, we're merely stating that historical and experimental sciences are not on any different footing, nor is one more scientific than the other, because truly there is no reasonable or rational justification for holding just about any view, it's just something that we humans do. This "makes all expectations to the future irrational (or unreasonable), even the expectation that we shall continue to feel expectation." Further, this is not even a "principle of probability . . . (since) all probable arguments assume this principle, and therefore it cannot itself be proved by any probable argument."[21] First, *mind truly blown*; second, it seems that time's arrow and time's cycle never did stand on a different footing, no matter what some high-minded experimentalists thought. At this point it is worth returning to Immanuel Kant, because Immanuel held that scientists actually derive knowledge not just by observing empirical phenomena but also by invoking a universal causal principle that we use to connect data with theory; whether those data are from experiments or from the analysis of the fossil record seems immaterial to us.

Note further that we humans don't typically run into brick walls, at least on purpose, or if we do run into them on purpose we're only going to do it once, unless we're really deranged or masochistic, because it bloody well hurts. (Where this puts stuntmen like Super Dave is another matter entirely, and beyond the scope of this or any essay.) Pay no mind to the fact that the universe is only 10 percent standard matter, with the remaining 90 percent comprising yet unknown substances like dark energy and dark matter: it's still going to hurt. Belief is never rational, yet Hume recognized that "the skeptic still continues to reason and believe, even though (they) assert that (they) cannot defend (their) reason by reason."[22] Or, if you're named Immanuel Kant, you can just invoke a "universal principle of causality."

From all these elegant interpretations Hume reached the conclusion that we need to stop thinking about the nature of things or knowledge so seriously and should just try to forget about the fact that reason doesn't actually exist. The absence of reason is like an annoying itch: to get it to stop, it's best to ignore it. Instead, those interested in science just need to get back to doing it, of course in a rigorous and principled way, and while using some solid ideas as a basic framework. Hume admonishes all scientists to not worry about extraneous mumbo jumbo. He went so far as recognize, brilliantly, that "there is no reason for studying philosophy . . . except that, to certain temperaments, this is an agreeable way of passing the time."[23] That is likely the real reason why Stephen Jay Gould wrote *Time's Arrow, Time's Cycle* and why Super Dave put on those false eyelashes after spending a brief time in a flaming domicile. It's certainly the reason we enjoyed the fruits of their labors.

———◆———

We've mentioned Kant, we've expounded on Hume, and we've quoted Bertrand Russell. Now it's time to wrap up this chapter. We didn't quite start out with a picture from an obscure seventeenth-century book by a theistically minded early geologist, but we hope we've demonstrated some of our bona fides. Unfortunately, neither Stephen Jay Gould nor Super Dave is alive to show us some love, but we hope they'd be pleased. Fortunately for us, writing this chapter was "an agreeable way of passing the time," and we truly hope you found reading it the same. We'd recommend that you listen to some Art Blakey and the Jazz Messengers before you dive into the next chapter. The song "Are You Real?" is a fitting tribute to all that is amazing about art and science, especially in this age of uncertainty, when many seem unsure of what "real" even means. That song is surely for real.

CHAPTER 5

Expanding Evolution

Organisms and Species, the Soma, and the Technosphere

Ever since biology became a profession, its first problem was to shed the supernatural as a source of explanation for the observable phenomena of life. Although there were many important forerunners, we date the emergence of evolutionary biology as a true profession with Jean-Baptiste Lamarck's 1801 publication on invertebrates, where he coined the term "fossil" and charted out an empirical/theoretical account of the origin of present-day marine species based on his discoveries of fossils in the sedimentary rocks of the Paris Basin.

We grew up hearing the same sort of smack about Lamarck that is still au courant: that he had a wacky idea about the "inheritance of acquired characters" as a proposed explanation for the manifestly interconnected diversity of life. With the coming of Darwin, it is supposed, Lamarck was blown out of the water, his notion replaced by Darwin's "natural selection." Seldom mentioned is Darwin's equally off-target preformationist notion of why organisms tend to resemble their parents and older generations.

Darwin perforce, and masterfully, treated inheritance as a "black box." "Grandchildren like grandfathers," Darwin observed as the first of three simple propositions that first spelled out the syllogism of natural selection in his *Notebook D*, the fourth collection of notes on transmutation, completed circa 1838. In other words, he was able to clearly formulate natural selection without having a clue why organisms indeed tend to resemble their progenitors. He simply recognized that this was so. The processes of heredity—and their implications for understanding evolution—didn't begin to emerge, arguably until the work of Auguste Weismann in the 1880s. Ernst Mayr said that after Darwin, Weismann was the most important evolutionist of the nineteenth century. Weismann's work was so significant because he dramatically expanded understanding of how hereditary information was passed down from generation to generation: that the basis of inheritance was via cells of the germ line. This work showed that evolution involving the inheritance of acquired characteristics, so called Lamarckian mechanisms of evolution, was no longer tenable. As such, Weismann became a pivotal early supporter of Darwin's ideas on natural selection.

Mayr overlooked Alfred Russel Wallace, the contemporaneous discoverer of natural selection, as well as the crucial role played by Lamarck in getting the ball rolling in the first place (not to mention the empirical catastrophist Georges Cuvier as Lamarck's foil, and the equally significant work of Giambattista Brocchi and other pre-Darwinians). Lamarck literally was a hero to Charles Robert Darwin, made especially clear in Darwin's late 1836-1837 *Transmutation Notebook B*, Niles's candidate for the single most important, *primus inter pares* document Darwin has left us. Entitled *Zoonomia*, *Notebook B* is an invocation of his grandfather Erasmus's book of the same name, a work that adumbrates serious discussion of evolution, called "transmutation" in those early days.

But it is Lamarck who shines and dominates Darwin's thinking in *Transmutation Notebook B*. At one point Darwin says that Lamarck was more important in 1836-37 than he was in 1809, Darwin's birth year and, more significantly, the publication date of Lamarck's *Zoologie philosophique*, far better known today than his crucial 1801 work on invertebrates. The point Darwin was driving home in his private recordings of his inmost scientific musings was that Lamarck insisted that life has evolved—and has done so by slow, steady gradual transformation of the features of species. Darwin got his gradualism not so much from Charles Lyell as directly from Lamarck. Lamarck was Darwin's hero, and not the Darwinian anti-hero we were taught to believe.

We could go on—and indeed Niles already has in his 2015 book *Eternal Ephemera*—but we'll add just one more point: Lamarck was playing his ideas off his great rival Cuvier, who insisted on stability and revolution as the hallmarks of the history of life, with its generation of diversity, the disappearance of species and entire higher taxa, and the appearance of new ones to take their place. "Take the place of" is perhaps the most resounding, repeated phrase in the entire evolutionary discourse in the first half of the nineteenth century. We believe it was 1830s-vintage Darwin, during the middle years of the *Beagle* voyage, who added geographic replacement of species to the concept of replacement in geological time: this in turn was derived from Lyell's 1832 dictum from the second volume of his *Principles of Geology*: as in space, so as in time. Cuvier never explicitly linked his take on the empirical patterns of stasis and change to a nonteleological, nonsupernatural, causal scientific explanation of the underlying causes that produced them. But the imagery of such "punctuated" patterns of various spatiotemporal dimensions is clearly rooted in his thought.

As a mirror of the atmosphere that still lives on in academe and much of the rest of life, the rivalry of Lamarck and Cuvier was in large measure rooted in the simple fact that they worked in the same place, the Jardin des Plantes in Paris. Lamarck rejected Cuvier's catastrophism—specifically, in strongly disbelieving Cuvier's pronouncement that species and even higher taxa have routinely become extinct during the rolling geological ages. Lamarck believed instead that species disappeared because they slowly transformed into descendants. He offered as evidence that 3 percent of the mollusks species he saw in what are now known to be the Eocene marine sedimentary rocks of the Paris Basin appear to be still living in the coastal waters of France. The rest, he asserted, had gradually changed to forms that dominate the molluscan fauna off modern shores. There were apparently no fossiliferous rocks intermediate in age known to Lamarck, so it was actually not so much of an empirical argument as an educated guess. But at least he advocated some sort of natural process underlying the appearance of the modern biota. Cuvier didn't.

We all know who, at least for a very long interval, won that argument. Darwin became aware of the reality of extinction through his experiences at Edinburgh under the tutelage of Robert Jameson, a great fan of Cuvier's. Darwin saw the evidence of geographic replacement of mammal and especially bird species below the Amazon Basin all the way down through Tierra del Fuego. He also saw the geographic replacement in exquisite microdetail on archipelagos, and so was aware that new species could and did emerge through isolation. In *Transmutation Notebook B*, Darwin demonstrates that he knows there is a natural process, as yet unformulated, that accounts for the changes in morphology and behavior in the course of "descent with modification"—but he says for now it must await discovery. He treated the process he later formulated as "natural selection" as a "black box" in *Notebook B*. He formulated natural selection in *Notebook D* a year or so later.

Darwin saw two possible sets of circumstances where the "black box" process of evolutionary change could be imagined to spring into action and produce palpable, observable evolutionary change. One was the microreplacement patterns

Mimus melanotis

Mimus macdonaldi

Mimus parvulus parvulus

Mimus parvulus barringtoni

5.1 Image of skins of various species of Galapagos mockingbirds of the genus *Mimulus* from the Yale University Peabody Museum of Natural History. Photography by and courtesy of Richard Prum, William Robert Cole Professor of Ornithology, Department of Ecology & Evolutionary Biology, Yale University and Curator of Ornithology, Peabody Museum of Natural History.

he saw across geographic space, and on islands in particular. He had a lot of his own data on that, and he thought the rheas of the pampas and Patagonian scrub, the foxes (now sometimes called wolves) of the Falklands (Malvinas) Islands and perhaps especially the mockingbirds of the Galápagos (figure 5.1), were eloquent testimony of the process which long after came to be known as geographic (allopatric) speciation. But he simply could not see how geographic isolation could occur over such vast stretches of land east of the Andes, though it was large land masses that held the highest diversity. And he was charmed and seduced by Lamarck's light-on-evidence, anti-Cuvieran vision of slow steady gradual evolutionary change. At least Lamarck was trying to propose a natural process to account for the diversity of life as we know it today.

INFORMATION AND THE THREE EVOLUTIONARY PATHWAYS

This is, however, *not* an essay primarily concerned with patterns of stasis and change in any system, including the original and still central subject matter of evolutionary biology: the origin of the forms alive we have with us today in horrifyingly shrinking numbers—which should be the main concern of each of us today and tomorrow.

What is at stake here is a deeper point: whatever the correct empirically based grasp of the spatiotemporal patterns of stasis and change, linearly gradual or episodic, may be, it is important that Darwin was able to specify natural selection without having anything resembling an accurate explanation for *why* organisms resemble their parents. But he knew that it was *something* that transmitted those instructions that made offspring look more like their moms, dads, and grandparents than, say, next-door neighbors (for the most part, anyway).

They didn't call it "information" back in the early and mid-1800s. We frankly do not know if Weismann used the German equivalent of the word "information." We'd love to know when the phrase "genetic information" first entered our professional lexicon. Could it have something to do with the advent of "information theory," dating back at least to World War II—coincidentally as the "modern synthesis" of evolutionary theory was beginning to be codified and emergent?

For in Darwin's black box on inheritance lay the as-yet unknown processual rules of transmission and the instructions on how to produce offspring that have that family resemblance. We feel confident that, if Darwin could simply read these last few sentences, he would agree that this was the nature of the contents of his black box.

The word "evolution" has of course long since become popular parlance. But it has also been applied rigorously to other sorts of systems beyond its original locus: the diversity of life. Languages are said to evolve. Likewise the gizmos we use to live our "civilized" lives have shown, we believe, a pattern in respects akin to biological evolution: the "material cultural implements" that have long since replaced significant biological evolutionary adaptation as an ongoing process to facilitate our existence, allow us to live long enough to reproduce, and hang on into increasingly older age (we are currently in our fifties and eighties, respectively, and it feels weird . . .). This realm of material culture has usefully been called the "technosphere." It is the realm of ancillary, or collateral, adaptations, begot in the human mind and instantiated in the things we make. And we know full well that the elements of the technosphere, too, "evolve."

We have learned about yet another system that is in fact an evolutionary realm in and of itself: cancer. Tumors don't just "grow." They evolve, as a subset of possible realms of somatic evolution—itself a foreign-sounding concept, but a real one—as somatic cells reproduce, making more cells of like kind. Processes of mutation, variation and winnowing selection occur in the cancer realm very much as they do in the evolution of adaptations in the larger-scale ecosystems of the world.

Niles had the privilege and pleasure of publishing a paper with cancer biologist James DeGregori in 2020 that explores the nature of cancer as an ecologically immersed evolutionary system. Cancer, of course, is a system that progresses and, with depressing regularity, can kill its host. What Niles brought to the table in this project is a take on technospheric evolution—and the thesis (that he has been expounding in one form or another since the late 1980s) that it is technospherically charged evolutionary events stemming from the agricultural revolution that have led to the current climate crisis and the Sixth Extinction, the present-day biodiversity crisis. It is driven almost solely through technospheric evolution. Thus we have two systems, not just one, where cancer (a form of somatic evolution) and material cultural evolution (technospheric evolution) both cause ecologically dysfunctional out-of-control population growth that, literally in the end, may be so destructive that it may kill off the host—precipitating their own demise.

And so there are at least three separate evolutionary domains: (1) "standard" biological evolution, mainly thought of in terms of sexually reproducing organisms; (2) somatic evolution (such as cancer, and immune system evolution); and (3) technospheric evolution, a subset of the more general phenomenon of the evolutionary growth of knowledge. A definition that embraces all three presents itself: *Evolution is the fate of transmissible information in an economic context.*

EVOLUTION IN THE TECHNOSPHERE: THE CORNET MICROCOSM

Niles developed an appreciation for the patterns and processes of technospheric evolution the old-fashioned way: empirically, through having amassed a considerable number of specimens from an evolving system and looking at the patterns of origination, stasis/change and death of specific designs throughout the history of a (more or less) discrete manufactured product—not trilobites this time, but rather cornets, which are brass wind musical instruments.

Niles's experience from the 1960s on up as a museum research scientist and curator of fossil invertebrates equipped him by the 1990s to apply very much the same sort of protocols he learned and followed with trilobites, to the collection, storage, and, eventually, analysis of "morphological" change in cornets. In addition to rocky outcrops, he haunted antique stores—and was in on the emergence and explosive growth of eBay, amassing a rich array of horns with (he thinks) a good sampling of all the major design innovations in two hundred years of cornet history. Niles's trilobite career had likewise introduced him to the realm of important collections around the world—and he visited the storied collections of most of the great musical instrument collections of the world, getting to know their curators, who further linked him to the vast web of amateur collectors. He also networked with fellow collectors worldwide.

Cornets are very like trumpets. We are speaking here strictly of piston-valved instruments—the sort that arose first in France and became the usual form of valved instruments of France and Belgium, the United States, the United Kingdom, and Canada and other elements of the nineteenth-century British empire. When Niles was a fourth grader in the 1950s he was loaned a school-owned Conn cornet. He was hooked—the cornet slept on the floor beside his bed when he first got it—and he has continued to play off and on ever since.

Cornets and trumpets are typically though not exclusively pitched in B♭, and in the 1950s they were considered musically interchangeable. But after some

5.2 The instruments pictured above are (top) Niles Eldredge's high school/college trumpet from the late 1950s or early 1960s, a knockoff copy by Rudy Muck of the more famous (and better) Vincent Bach trumpets, themselves a knockoff of the more famous (and better) French Bessons of the 1920s—themselves only slightly altered from the patent drawings of 1871. Imitation, the sincerest form of flattery, is a common phenomenon in the transmission and evolution of the elements of the technosphere. Below Niles's trumpet is a contemporaneous Rudy Muck cornet, itself a knockoff of the evolved cornet form of the early to mid-twentieth century cornet.

resistance Niles came to realize the truth of the proposition that cornets and trumpets had different "origin stories"—different beginnings for different purposes. The similarity in sound and even in shape ("morphology") between cornets and trumpets since the early to mid-twentieth century is really a matter of deliberate, intentional convergent evolution (figure 5.2).

What matters most in the present context is that cornets are stunningly diverse in terms at least of their body shape. Trumpets, on the other hand, are monotonously, almost boringly, all pretty much the same. They have been in virtual stasis since their existence—in the form of shop drawings published by Florentine Besson in her 1867/1871/1872 French (simultaneously in Belgium and the United Kingdom) patent (trumpets were added in 1871). It was her granddaughter

Meha who produced what were widely considered the very best quality trumpets, still in Paris, in the years immediately after World War I and continuing for a decade until the worldwide depression drove the Besson firm (along with so many other commercial ventures) into bankruptcy in 1931, the cultural equivalent of being hit by a meteorite. Meha became known as the second "Mme. F. Besson" in a marketing effort to dissociate the company from her scoundrel father, Adolphe Fontaine, who had tried to make "F. Besson" stand for "Fontaine Besson." Together with Florentine's daughter Marthe, Meha's mother, the firm was run solely by these three women after founder Gustave Besson transferred it to Florentine in 1858 after establishing a factory in London in addition to the one in Paris.

Piston-valved trumpets, no matter where, when and by whom, have hardly changed in overall configuration since 1871. Still, the 1920s was the decade when Louis Armstrong famously switched from cornet to trumpet, symbolizing if not actually causing the switch from cornets to trumpets. Armstrong evidently had learned about trumpet use in New York City when he played briefly with the Fletcher Henderson Orchestra in 1924–25. For reasons of style, sound projection, and perhaps even studio recording use, trumpets emerged from their little-used pre–World War I status to the soprano brass instruments of choice. The change is reflected in the first appearance of trumpets in the Sears Roebuck catalogues around 1924, and it appears as well in the Google Ngram Niles obtained by juxtaposing "cornet" and "trumpet" in figure 5.3, where abrupt changes in relative citations clearly began just after World War I.

So, to return to science for a moment, cornets seem analogous to groups in the biological realm that display high rates of speciation, like trilobites; by contrast, trumpets seem analogous to biological groups that show low rates of speciation, like worms. These are respectively termed "high" and "low" volatility groups; we will expound on the concept of volatility and its relevance to evolution in the next chapter.

But back to cornets and their profligately profuse morphological variation—and history so very different from that of the B♭ trumpet. Niles swears much of the evolution of these instruments was driven by canons of stylistic beauty that had little or nothing to do with how well they played or how good they sounded. It is perhaps no coincidence that the first western-world "rock star" was the cornet virtuoso Hermann Koenig, a German-born musician who drew large crowds in France and the United States but was especially a fixture of London's Covent Garden in the 1850s.

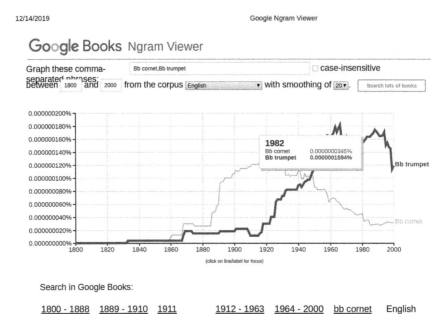

5.3 A Google Ngram that shows how the number of references to "cornet" and "trumpet" in published documents has changed through time.

Naturally enough there were periods of stasis and change, with a little seeming protracted gradual change thrown in for good measure, in the evolutionary history of cornet design. There were extinction events followed by evolutionary change that incorporated the products of many makers nearly simultaneously, reminiscent of Elisabeth Vrba's turnover pulse events (figure 5.4) that regularly punctuated the history of life at least since metazoans hit the scene half a billion years ago (and discussed in greater detail in chapter 8). And it was cornets that converged, evolutionarily speaking, on trumpets—with longer bells and shallower mouthpieces—leading to the situation of Niles's youth, where the two looked similar and sounded much the same, especially in inexperienced hands.

Niles would be tempted to refer to the technosphere as Richard Dawkins's "extended phenotype" were it not for the fact that Richard enjoins his readers in his eponymous 1982 book to employ the term only when there is direct linkage to the human genome. Bruce is less sure if Dawkins even understood this distinction himself.

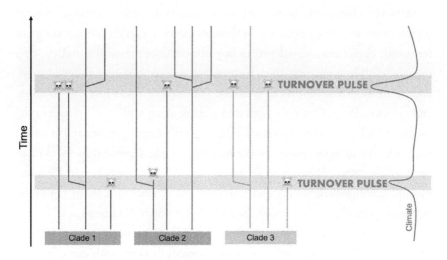

5.4 Diagram illustrating Elisabeth Vrba's turnover pulse hypothesis. Courtesy of Jonathan Hendricks, Paleontological Research Institution, and the *Digital Encyclopedia of Ancient Life*.

The flagrant purposefulness of technospheric evolution is perhaps its canonical feature. Gizmos are ancillary, collateral adaptations—human-conceived and human-made devices that serve purposes—mostly economic, but sometimes reproductive (e.g., in vitro fertilization).

More apt, of course, is Dawkins's earlier now-viral term "meme," co-opted for an unintended use on the internet. Niles has shied away from using the word "meme," so clearly intended to be a particle of information analogous to "gene," for the simple reason that a gene is itself a corporeal entity that carries a piece of "genetic information"—and memes are said to be pieces of memetic information without a comparable instantiation in physical reality. Perhaps this is a cavil or merely a distinction without much of a difference, but though Niles likes the term "memetic information," he cannot bring himself to employ the term "meme" itself. (Bruce had forgotten that Dawkins invented the term "meme." He cannot bring himself to use the word now because he is extremely disaffected by the fact that Richard Dawkins has sold more than two million copies of *The Selfish Gene* even though prodigious aspects of the book have no scientific validity. Thus he tries to forget most things associated with Dawkins, except for the *South Park* episodes that featured him.)

Memetic information is indeed another matter. Humans exchange cultural information—memetic information—through speech and oral tradition, the written word, depictions, actual preexisting objects (like measuring and copying someone else's instruments, which frequently happened in cornet design history), and via many different sorts of symbolic systems (Niles was listening to music and occasionally picking up a horn and playing along with it as he wrote this) used with each other, from parents to offspring, offspring to parents, and of course with other contemporaneous, in-the-moment people. Lateral transfer is more common than vertical, but both are important. We even mine the past for information, in focused studies (e.g. "what did Darwin say about isolation?") or in the fading practice of simply cruising the shelves of a wonderful library.

Niles learned about cornets from the specimens themselves (Bruce learned about cornets from Niles), many of which mercifully come with one of an historically ordered sequence of serial numbers. (Fossils don't have serial numbers, though their relative position in time is usually clear from their occurrence within sequences of layered rocks.) Cornets can be measured, their parts separated and inspected, their innards revealed by x-ray and other means. People have written about cornets, as figure 5.3 shows. There are patents, replete with statements of intent and illustrations. There are preserved shop mandrels used by makers—the physical model which makers used to make bells and parts that are consistently the same from one horn to another in the day-to-day, year-to-year consistency in production of horns of a particular model in a particular maker's shop or factory. There is advertising (often more informative than the horn itself!). There are personal accounts, photographs, and all sorts of memorabilia. All this information is usefully thought of as memetic, copyable in its transference.

Much of the early history of cornet making undoubtedly involved trial and error: for example, what the proper taper of the tubing must be to make a horn that plays with a particular sound and is internally in tune up and down the scale. That would be an experimental form of selection. Niles has seen many patents where the horn, or newly configured valves, or tubing plans never seem to have been put into production: selected as unfeasible for a host of possible reasons, including chance alone, difficulty in manufacturing, being deemed an insufficient improvement to warrant new machinery and production protocols, and so forth.

Then there is the selection/winnowing of products in the marketplace: which design produces the better results, whether in facility of play, beauty of tone, fidelity of intonation, or simple attractive design. Better-selling models (also in part determined, of course, by price and extent of advertising and distribution) are nearly always copied by rival makers. Selection in technospheric evolution comes in many forms, at different levels arising with the makers, and then by their customers and the exigencies of larger-scale economic forces. There were no brass instruments made for several years during World War II in the United States because every scrap of brass (probably including old, recycled brass instruments) was turned into shell casings.

Unlike biological and somatic evolution, variation and selection are in the main purposeful, based on the fate of transmissible ideas emanating from the human brain—almost always a single person. But Niles will say that one of the most-used trumpet models in jazz from the 1930s through the 1970s was the Martin Committee. Roy Eldridge, Dizzy Gillespie, Miles Davis, Chet Baker, and legions of others played them throughout most of their careers; the horn was indeed designed by a committee of skilled craftsmen (a useful reminder that it isn't *always* true that committees never produce anything of value). Not only were they good horns, but one of the last of Davis's Martin Committees (by that time from a later maker) was auctioned off in late 2019 for more than $200,000.

MAJOR EVENTS IN THE EVOLUTION OF CORNET DESIGN

In the first decade of the twenty-first century, designer Bruce Hannah and Niles made a presentation on the evolution of design to a salon-style gathering at the Museum of Modern Art in New York City. Part of their discussion centered on a short PowerPoint presentation that Niles compiled and annotated on cornet evolutionary history. Blowing the digital dust off, we bring some parts of it to light here, with a few further comments, to bring out what has emerged as some of the general features of technospheric evolution (figure 5.5).

To produce a sound in a piece of cylindrical/conical piping (even a garden hose!), one way to go is to devise and insert a mouthpiece of the right diameter so that the lips can produce a "buzz." Only the upper lip vibrates. The tubing acts as a resonator and produces a clear sound, with the pitch determined primarily by the length of the tube (the "horn") and to some extent by the internal width of the tubing. A horn roughly four and a half feet long is pitched in concert B♭ (the conventional modern trumpet and cornet pitch). A given length of tubing also constrains the available notes to those dictated by acoustical physics to the

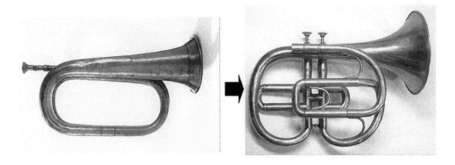

5.5 Image showing the valves added to a natural horn circa 1820.

"overtone series," a fundamental lowest note (often difficult to produce and not clear sounding—a "pedal" tone). The next higher note is an octave up. If the fundamental and the next higher up is a C, it is followed in ascending series: G, then C, E, G, B♭, C ("high C"), then D, E, F, and G. On lip-blown instruments, progressively higher tones are achieved by increasing the frequency of the lip-produced buzz. The higher you play, the closer the notes become. Baroque trumpets lacked such contrivances, so that only virtuosos could play Bach, Handel, Purcell, and the like in the extreme high register. We still know the name of Bach's favorite trumpeter: Gottfried Reiche, whose portrait still exists, so special were those players.

Western music uses a twelve-tone scale—so the lower notes cannot be used to produce a simple tune, such as Thelonious Monk's "*Straight, No Chaser*," a particularly brilliant version of which can be found on Miles Davis's album *Milestones*. Rather, only the predetermined notes of the overtone series can be used in instruments such as the bugle on the left of figure 5.5. The length of the tubing needs to be changed to get those missing tones—by using tone holes (with or without keys, like a saxophone) or a slide (like a trombone) or some sort of valve that can redirect the windway into added lengths of tubing to increase the total length of the tubing.

All three were tried, and marketed, with lip-blown brasswinds and their progenitors. By the early 1800s, metallurgical engineering is thought to have advanced sufficiently to enable the production of valves sufficiently tight to avoid leakage, the kiss of death to decent brass sound. Used in combination, valves enable production of all twelve tones on the Western scale, especially in the lower and middle registers, where some of us think the sound is the most beautiful.

It takes three valves to get all twelve tones. It's thus a mystery to Niles why the first known design of a valved soprano instrument—a cornet circa 1825—had only two valves. Niles thinks, given later developments, that the simple symmetrical beauty of the two-valved version may have been a factor. Piston valves as opposed to the still-in-use rotary valves preferred in Germany and elsewhere were the choice in Paris, at the outset the kind designed, ironically enough, by a German: Heinrich Stölzel.

Science enabled a cultural yearning that has been around probably as long as humans have been singing—a need to transform whatever technospheric invention ever produced to be able to produce as many notes as our human voices. It took only four years for the third valve to be patented (figure 5.6). With the third valve quickly added, the emulation of the chromatic scale of the human voice had been achieved, at least as far as Western ears were concerned. Function drove design in this case.

The original designer of the two-valved model is generally thought to have been the Parisian Halari around 1825. François Périnet, also Parisian, added the third valve in 1829, though many other makers were producing copies of both right away. Though less harmonically functional, two-valved cornets never truly went extinct, though the presence of two-valved instruments of various pitches in marching bands seems to be an example of the "Lazarus effect" by which underlying concepts in the technosphere can be resurrected at will, whether by deliberate plan or by the mere idea that two valves (cheaper to make) serve a particular purpose just as well.

5.6 Image showing the addition of a third valve circa 1830.

5.7 Image showing the addition of the modern valve design in 1842.

Next came a major improvement, a new valve design (figure 5.7). The original Stölzel valve was constructed with the windway running up inside the valve (note the main tubing running up into the last valve at the bottom in the three-valved cornet on the left). This meant that the valve could be no wider than the diameter of the tubing connected to it, a major design constraint that among other things exposed the moving valves to extra wear and deformation simply through the distorting pressure generated by the left-hand grip of the performer.

A decade after the third Stölzel valve had been added and adopted by a number of makers in Paris and elsewhere, Périnet invented a new form of piston valve that decoupled the wind way from the vertical interior of the Stölzel valve. This Périnet valve is still in use today, in its original form, or, most commonly, in the form invented by Mme. Besson in her 1870s patent. Through the years, as well, different, harder metals have been used instead of the usual brass of the original versions, also to make the valves less susceptible to wear. And decoupling the windway from the vertical interior of the valve meant that it could be built with thicker, sturdier walls, preventing deformation by the player's grip.

This is "evolution" in a sense. Both valves are similar piston valves. The fingering of the notes remains identical with both valves. But in another sense, beyond being a piston valve with the same mechanical and musical function, the Périnet was an alternative, clearly superior design to the Stölzel. It wasn't until the late 1990s that Niles, looking at an image of a cornet in a Parisian friend's published catalogue, recognized that his oldest Adolphe Sax cornet, built in around 1842 in

Paris, was not just another three-valved cornopean, the common term for older cornets equipped with Stölzel valves. None of the cornet's tubing was connected to the vertical interior of the valves. Instead they were the original form of Périnet's valve replacing the Stölzel valves, but done so cleverly that the original form of the three-valved cornet was faithfully retained.

It was like seeing an early experimental car, looking for all the world like a buckboard with the horse removed, replaced by an electric, steam, or internal combustion engine with a drive shaft and a mechanism to steer the contraption down the road. Such a design event is sudden, and its hallmark is human intentionality. The "evolution" is more like directed variation, sudden and either/or: there are no intergradations of the conventional sort seen in biological evolution, even in most species-level events involving a "punctuated" pattern, where change happens fast in comparison with the much longer period of stasis before and after (see chapter 2).

This "better mousetrap" pattern of replacement design is utterly characteristic of the way novelty is introduced in technospheric evolution. Patents, intended to preserve ownership of a particular design (though the logic of being able to patent a natural gene sequence is beyond comprehension) are famous for spurring on innovation by those who want to enter the market for a particular sort of gizmo. Patents are intended to thwart or at least delay the rampant phenomenon of lateral "theft of idea" form of information transmission that lies at the heart of human cultural evolution generally. Sometimes designs are licensed to other makers. And sometimes new ideas are taught or otherwise made public through publications including advertising. Niles has known cases of cornet and trumpet makers who routinely measure and dissect the bodies of the instruments of their competitors to try to find out what the design secrets are that make their products sound or even just look so good.

But other such intentionally designed replacement events seem simply to be improvements, where the same basic function, the same use for a mechanical ancillary adaptation, is achieved in a starkly different way. Thus the "Hannah principle." Bruce Hannah, Niles's friend and partner in the MOMA design salon experience, first pointed out this punctuated pattern to Niles, involving intentional redesign rather than a smoothly gradual derivation of structure and form as a major hallmark of technospheric evolution. A different way of solving the same functional problem. A better mousetrap.

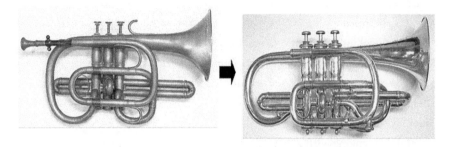

5.8 Image showing how in 1855 the bell moved to the left of the valves, producing the iconic cornet design.

Once the engine got onto the buckboard, the door was opened for all kinds of changes in overall body design of cars. Same was true of cornets as they entered their golden age in the mid- to late 1800s (figure 5.8).

Cornets in their earliest years all had the external windway tubing to the right side of the piston valves—very much as if the addition of valves to a posthorn/bugle to fashion a cornet in the first place was just a matter of slapping a set of valves onto the side of the bugle. There are two distinct external tubes: one conducts the windway from the mouthpiece to the valve system, the other takes it from the valves and emits it through the bell flare and out into the world. All these brasswinds are simply resonators that convert a buzzing, disturbed air column into the dulcet (or sometimes raucous) sounds these things can make.

In around 1855, the French maker Antoine Courtois, most successful of an entire family of instrument makers, introduced a *nouveau modèle* where the bell branch was shifted to the left side of the valves. The silver-plated cornet on the right of figure 5.8 was made by Courtois in 1855 and is the oldest example of a bell-on-the-left piston valved brass instrument yet found, at least as far as Niles knows.

Resisted for decades by Courtois's native French customers, this design was an immediate hit in London and elsewhere in the British Isles. It became the standard design of all subsequent Périnet-valved cornets and trumpets, spreading to the United States by the 1870s and (grudgingly) adopted as well in its native France as the nineteenth century wore on. They became known, unsurprisingly, as "English bell" models.

Why the change, sandwiching the valves between the bell tube, now on the left side, and the mouthpiece tube ("leadpipe"), still on the right? Though never stated by Courtois or anyone else, the simple change made the horn undeniably easier to grip and to hold steady while playing. Simple as that.

The introduction of Périnet valves in the 1840s did not spell the extinction of the older Stölzel valved cornets: they lived on as the cheaper "student" models for decades—especially in France. By the 1860s and 1870s some of these Stölzel valved cornets were retrofitted with the newfangled English bell design. In other words, there is a great deal of freedom in the usage of various design concepts, where new ideas can be used on a wider range of older models that were not strictly speaking the proximate "ancestors" of the latest design.

Although French makers would still build an old-fashioned French model cornet up at least through the 1930s, it was the English bell Courtois-style design that carried the day as the iconic cornet of halcyon days of the Victorian era. The Courtois instruments were the choice of most solo artists, but there were literally dozens if not hundreds of imitators. Some were good enough to compete with better quality horns, but others were cheap knockoffs that flooded the markets of the Western world. External design alone does not guarantee a quality product. At the turn of the century there was something of a "turnover event" where production of the Courtois style Victorian style cornet dropped, in part (Niles thinks) because they weren't fashionably new enough to still be suitable for the new century.

The trend to make cornets longer and more like trumpets was also setting in, particularly in the United States in the years before World War I. The tipping point, the nearly wholesale switch from already "trumpetized" cornets to actual trumpets, came right after World War I, as production again picked up in Europe and North America.

As a final word in this consideration of the empirical details of the evolutionary history of cornets, revivals are rampant in human-designed physical systems: nostalgia may be one of the sappier ingredients of the human emotional palette, but it can and has been an explicit component in design evolution. The single most stable feature of the Victorian cornets was the deep bend at the rear of the bell branch—the aptly named "shepherd's crook." A holdover from the very first cornets ever built in the 1820s, shepherd's crook cornets all but disappeared everywhere in the twentieth century, as the cornet depicted along with the trumpet in figure 5.2 reveals. They survived mostly in British brass bands. But a wave of nostalgia for these short, decidedly untrumpetlike horns swept the marketplace in the final decades of the twentieth century, carrying on

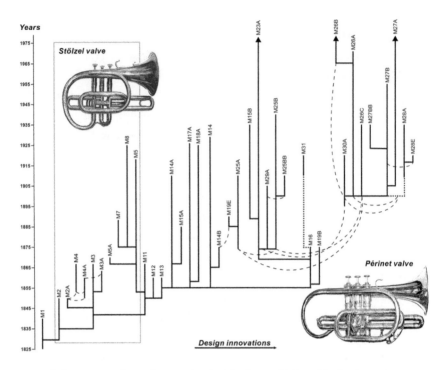

5.9 A phylogenetic analysis of cornets by Ilya Tëmkin and Niles Eldredge. Diagram provided by Ilya Tëmkin and used with permission of the journal *Current Anthropology*, University of Chicago Press.

to the present day. This "Lazarus effect" is ample evidence that design information does not die with its disappearance in the marketplace (in the technosphere, extinct may not be forever, unlike in biology). Makers can and do make old-fashioned replicas of all sorts of things: think of radios looking like their conceptual ancestors from the 1920s, powered, in true Hannah principle fashion, now by electronic circuitry vastly different from the original vacuum tubes.

Figure 5.9 shows the results of a phylogenetic analysis of cornets by evolutionary biologist Ilya Tëmkin of Northern Virginia Community College and Niles, using a database of cornets Niles created.[1] The vertical lines represent the known temporal distribution of discrete cornet models. The horizontal dark lines depict patterns of evolutionary connectedness. The lighter dashed lines depict some of the movement of design elements outside the ancestor/descendant analytic scheme (here, sister-group monophyly).

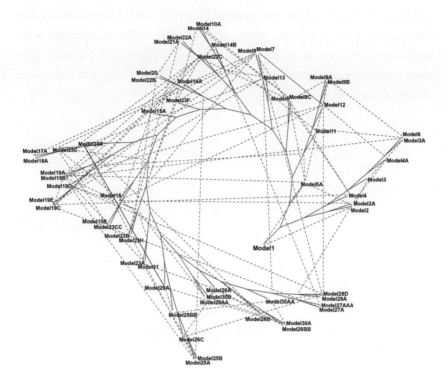

5.10 Reticulogram by Ilya Tëmkin and Niles Eldredge showing the connections between cornet models. Diagram provided by Ilya Tëmkin and used with permission of the journal *Current Anthropology*, University of Chicago Press.

By contrast, figure 5.10 is a "reticulogram" of the information connectivity between cornet models latent in the same database used in the phylogenetic analysis.

Figure 5.10 looks more like some diagrams of bacterial relationships involving tremendous amounts of horizontal gene transfer than anything known about relationships among a group of trilobites. More work is needed in this area to consider the best ways to analyze such systems and better quantify the relative proportions of vertical descent and lateral transfer in technospheric evolutionary systems.

Intentionality of design and rampant lateral transfer of information are two of the main characteristic features of evolution in the technosphere—and in cultural evolution more generally. The information originates and is maintained ultimately in the minds of the designers/makers, their contemporaries and descendants—even though it is preserved in symbolic form on paper and other media—and of course in the manufactured objects themselves. Humans make more humans; cancer cells make more cancer cells. But cornets don't make more cornets. Only humans can do that.

CHAPTER 6

Declining Volatility

A General Evolutionary Principle and Its Relevance
to Fossils, Stocks, and Stars

T here are certain things once common that are no longer so. Liberal
Republican politicians, like former New York City Mayor John Lindsay,
talented actors in TV sci-fi series, like the great William Shatner, and
the glorious trilobites (figure 6.1). The same holds for ammonites as well (figure 6.2).

Why are certain groups like trilobites and ammonites entirely extinct? That
may be easier to answer than the reasons behind the disappearance of certain
types of politicians and recent incarnations of Will Shatner, which are beyond
our comprehension.

VOLATILITY, EVOLUTION,
AND THE EXTINCTION OF SPECIES

The explanation as to why trilobites and ammonites are no longer encountered,
except as fossils, relates to the fact that both are evolutionarily volatile groups.
Volatility allows us to connect the extinction of trilobites to the behavior of stocks
in the market and stars in the universe. This relationship was explored in a 2013
paper in the journal *Palaeontology* by Bruce and his colleague Adrian Melott, an
emeritus professor in the Department of Physics & Astronomy at Bruce's home
institution, the University of Kansas. Because volatility is an important concept,
it behooves us to define it. What volatility means in an evolutionary sense is a
high rate of speciation and extinction, such that volatile groups are those that
have relatively high rates of evolution. What volatility means in the context of
stocks is that their price tends to move up or down at a greater rate relative to
the rest of the stock market. Finally, when it comes to stars, volatile stars are those

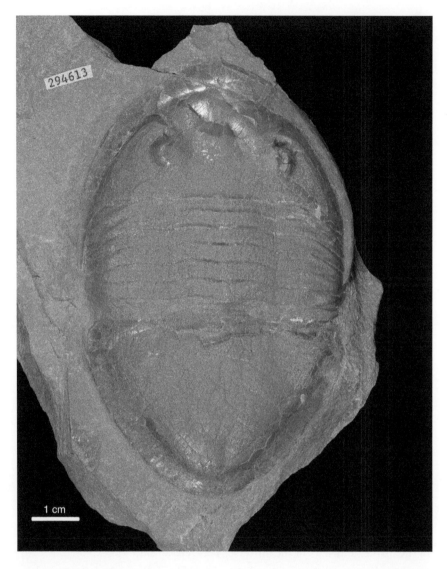

6.1 The trilobite *Isotelus iowensis* from Pike County, Missouri, KUMIP 294613. Photo by Natalia López Carranza (KU).

that form and explode at a rapid rate. A key aspect of volatile entities, be they species, stocks, or stars, is that over the long term they are not stable. Thus, they exhibit a high rate of turnover and do not persist.

Several paleontologists, including Norman Gilinsky and David Raup, previously noted that volatility may explain why certain types of animal groups

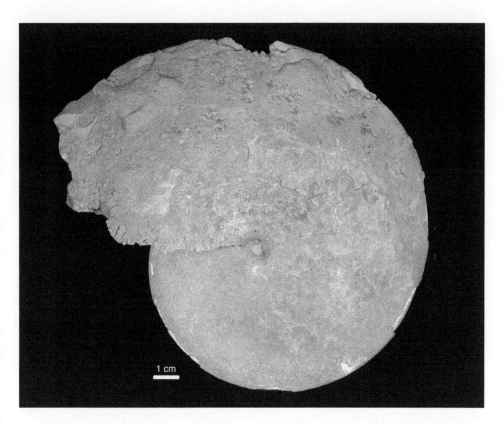

6.2 The ammonite *Collignoniceras springeri* from Sandoval County, New Mexico, KUMIP 150808. Photo by Natalia López Carranza (KU).

persist over the long term, while others do not. This is related to the fact that groups that have a high rate of speciation also tend to have a high rate of extinction. That in turn is a phenomenon first identified by Niles, as well as by paleontologists Steve Stanley and Elisabeth Vrba. The reason for the association between speciation and extinction is that similar processes cause speciation and extinction. For instance, one important process relates to the size of the area a species occupies (its geographic range). Species that are restricted to narrow geographic ranges are more likely to go extinct, but they are also more likely to speciate. Let's consider the extinction aspect first. It holds true for two reasons: first, being confined to a narrow range typically implies rarity, and something rare is more at risk of disappearance than something common; second, a narrow range means that one local event could wipe a species out. Nobody worries about the Norwegian brown rat going extinct, because despite its name it occurs everywhere (not just in Norway) and is incredibly abundant. By contrast, that butterfly species found on a single mountaintop in West Virginia is of concern to conservationists. Now let's consider the enhanced speciation aspect of species with narrower geographic ranges. These species are more likely to speciate because they tend to contain populations that are more likely to become geographically isolated from one another, and geographic isolation is a key factor that leads to evolutionary divergence and eventually speciation, as we will see.

Another aspect of species with narrow geographic ranges is that they tend to be ecologically specialized (the technical term used for these is *stenotopes*), depend on one or a few prey items, and are able to tolerate only a few environments. By contrast, species with broad geographic ranges tend to be ecological generalists (the technical term used for these is *eurytopes*) that can consume many different types of prey and/or live in a plethora of habitats. The stenotopes perforce tend to be confined to narrower places, while the eurytopes are more broadly distributed, and thus the stenotopes, because of their ecological and/or environmental specialization are more susceptible to extinction and more prone to speciation.

Now, consider what happens to volatility through time. This figure is from Bruce and Melott's 2013 paper "Declining Volatility, a General Property of Disparate Systems" (figure 6.3). (In this figure, time is on the horizontal or x-axis and is arranged from left to right, with very old on the left and present day at the far right.) The pattern is quite clear and highly statistically significant: when the first abundant animals appear in the fossil record, which was roughly 550 million years ago (mya), the volatility was quite high. And it has been declining ever since. Indeed, the drop in volatility over the ~100-million-year interval of animal evolution (and extinction) between ~550 and 450 mya is particularly precipitous, with basically the bottom falling out of volatility; levels of volatility never recovered. These data are based on the relative origination and extinction rates of marine animal groups in the fossil record, and what they mean is that in an important respect life is much more boring evolutionarily today than it used to be. And by "boring" we mean that it evolves and goes extinct at a much less explosive rate.

It might even seem counterintuitive, but in fact it is groups with high rates of evolution and speciation that are particularly at peril. High rates of speciation are perilous for two reasons. First, high rates of speciation are correlated with high rates of extinction, and second, once a group's diversity reaches a zero point, which is much more likely to happen if a group has a high rate of extinction, then

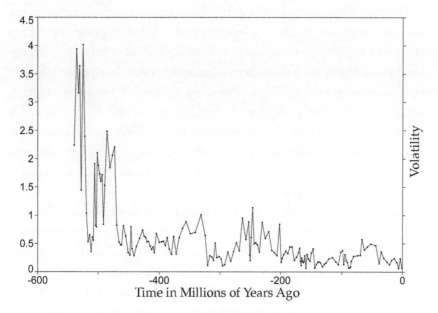

6.3 The change in the volatility of marine animal genera through time in the fossil record, modified from Bruce Lieberman and Adrian Melott, "Declining Volatility, a General Property of Disparate Systems: From Fossils, to Stocks, to the Stars," *Palaeontology* 56 (2013): 1297-1304, used with permission of John Wiley & Sons and the journal *Palaeontology*.

there is no possibility of recovery: once you're gone, you're gone, like the passenger pigeon. This is perfectly expressed in the fundamental admonition "extinct is forever" from the field of conservation biology.

A group with high volatility and concomitant high extinction probability is especially at risk of reaching a zero-diversity point, and thus vanishing entirely, during times of mass extinction. Now, as we've previously mentioned, mass extinctions are times when the extinction probability climbed dramatically for all life forms; there have been five major ones in the history of animal life. These represent times over the last ~500 million years when 60 percent to perhaps more than 95 percent of all marine animal species went extinct. (A significant number of terrestrial plant and animal species went extinct as well in most cases, though the percentage values are not quite as well understood.) These extinction events were truly global in scope and, depending on the individual mass extinction, may have transpired over the course of a few years to a few hundred thousand years. (For some additional discussions on mass extinctions see chapters 7 and 8.)

However, irrespective of cause and duration, mass extinctions matter for our discussion here solely due to their inherent mathematical properties. Consider a group with high volatility and thus an innately high rate of extinction; then imagine that conditions get bad, as they do during a mass extinction, such that the extinction level is dramatically elevated across the board. When a group containing species with a high probability of extinction is confronted by a huge uptick in the rate of extinction, the net result is a tremendously high probability of extinction for all its component species. That means there is a great increase in the likelihood that the group's diversity will fall to zero, an evolutionary point of no return, and the group will disappear.

Now consider the trilobites. For most of their evolutionary history they were doing pretty well in the sense that their diversity was rising, or at least it was stable and not falling. However, during times of mass extinction they happened to have fared very poorly. For instance, they suffered a major hit to their diversity during the first, late Ordovician mass extinction. Their diversity rebounds somewhat, but then they are hit again very hard during the crisis of the Late Devonian. Indeed, all but one of the major groups of trilobites, the Proetida (figure 6.4), disappeared.

After the end of the Devonian trilobite diversity only rebounded slightly, and it stayed at a low level for the next roughly 100 million years. Then the proetids, and thus all trilobites, disappeared during the end Permian mass extinction, what Stephen Jay Gould called the "granddaddy of all extinctions." The granddaddy term seems to derive from the media's branding of football's Rose Bowl, which is the granddaddy of them all because it was the first. The Permian isn't the first, but it's the worst. Steve's granddaddy appellation here may relate to the fact that because of its enormous effects in transforming the biota it may have been one of the first, if not *the* first, identified in the fossil record. Or he simply may have been giving a shoutout as a sports fan to the dulcet play-by-play stylings of noted sportscaster Keith Jackson, who first dubbed the Rose Bowl "the granddaddy of them all."

Scientists have always had trouble coming up with an explanation as to why the trilobites vanished. Perhaps the most traditional explanation relied on invoking the standard Darwinian principle of natural selection, which posited that trilobites were slowly but inexorably driven to extinction over the course of many hundreds of millions of years by competition from crabs or other crustaceans.

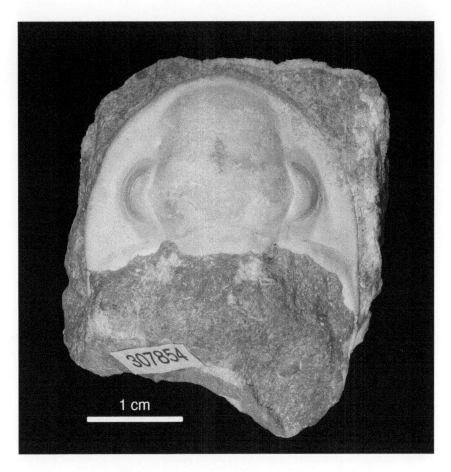

6.4 The head of the proetid trilobite *Ameura major* from Wyandotte County, Kansas, KUMIP 307854. Photo by Natalia López Carranza (KU).

This explanation was invoked in a gestalt sense because trilobites and crustaceans generally resemble one another, they are both arthropods after all, and trilobites lived in many of the same habitats that crustaceans did and do. However, the explanation is lacking for a variety of reasons, including the fact that if crustacea were outcompeting trilobites they would have driven them extinct over the course of a few thousands of years, not hundreds of millions of years, because competition is such a powerful evolutionary force. We are not saying that natural selection does not operate in the case of trilobites or crustacea or anything else: it simply must operate because it is a statistical law. Instead, we are arguing that the extinction of trilobites is an inherent byproduct of mathe-matical principles. Probabilities of extinction happen to differ across major groups of animals, and the history of life is punctuated by very rare but stupen-dously massive catastrophes that elevate probabilities of extinction universally. These in turn most dangerously elevate probabilities of extinction in species with already high probabilities of extinction. Thus, large numbers of species across the board, but especially of a certain type, such as trilobites, get elimi-nated quickly and can never repopulate Earth.

We can tell a story similar in broad brush strokes for the ammonites (for the professional paleontologists keeping score out there, it would be more technically correct to use the term "ammonoids"). The ammonites first appeared during the Devonian period, shortly before the Late Devonian crisis. They scraped through that crisis, and then their diversity rose substantially. In terms of overall diver-sity things were going great for the ammonites until they took a tremendous hit at the end Permian mass extinction. Very few species remained after this deba-cle: they seemingly just squeaked through again.

Well, good times at least in terms of taxonomic diversity, for the ensuing ~50 million years of the Triassic period. But then yet again ammonite diversity gets smacked down, thanks to the late Triassic mass extinction. Then all seems to be smooth sailing for the next ~135 million years or so and the number of ammo-nite species rises substantially until, all of a sudden, a giant rock falls out of the sky and precipitates the end Cretaceous mass extinction, when the curtain closes on the final act of the ammonites. They disappeared along with many other groups, including, most famously, the large nonflying dinosaurs. Our good friend and colleague Neil Landman of the American Museum of Natural His-tory has beautifully teased apart the final death spiral of this planispiral group: they seem to have briefly survived for a few scant centuries after the end Creta-ceous event, in, of all places, New Jersey.

From the perspective of life, in the long term it pays to have low evolutionary volatility, if survival is a good thing, and in the game of Darwinian evolution it most definitely is a very good thing. This volatility is sadly something that no group of species, or even an individual species, gets to choose. Instead, it's a prob-ability determined by various parameters of species and the organisms they con-tain. Perhaps the expectation might have been to suppose that groups with high rates of evolution and speciation were the most likely to survive in the long term, because maybe they would be the ones most likely to adapt in the face of changing circumstances. However, because a high speciation rate is associated with a high extinction rate and the existence of punctuated equilibria, high rates of evolution are not actually conducive to a group's long-term persistence. It is a sobering thought that humans are of course mammals and mammals are in turn a group with relatively high speciation (and extinction rates), given that we are in the midst of the sixth mass extinction event, the Biodiversity Crisis, triggered by our own species' activities. These activities are thus placing us (not to mention all other kinds of species) in substantial peril.

What volatility means is that, in the long term, the best evolutionary strategy for a clade (not that it gets to pick a strategy) is to have a very low rate of evolution and speciation, because of the concomitant low rate of extinction it usually affords. This shields them from the buffeting effects of dire extinction events. If we consider animal groups that may have this so-called best strategy, it is the annelids, or worms. When it comes to long-term duration of its constituent clade, it's good to be a worm.

VOLATILITY AND MATHEMATICAL EXPLANATIONS OF EVOLUTION

Views like these would very much resonate with those that Steve put forth throughout his career. For instance, he repeatedly tried to emphasize what he saw as challenges to or inadequacies of the standard treatment of evolution, sometimes referred to as the neo-Darwinian paradigm. Steve held that proponents of this neo-Darwinian view adhered too rigidly to the notion that all evolutionary change was adaptive and related to competitive interactions among different individuals; he lamented that important structural principles and inherent physical or mathematical properties were ignored. If something in biology or paleontology could be explained by basic math rather than by recourse to invoking biological processes, Steve would usually be all in. This type of approach was highlighted in his own work on the Irish elk, whose appellation is somewhat of a misnomer because it isn't confined to Ireland but instead occurs in places as far afield as Kansas. The Irish elk was known for its excessive size, by the standards of artiodactyls (the family that includes elk and other varieties of deer), and especially its tremendous rack of antlers. Because the antlers in this extinct species were so large, traditionally a variety of adaptive explanations were put forward to explain how and why the antlers in this species would have grown so big: these usually focused on the putative use of antlers in battles among males to impress females of the species, with resultant selection for larger antlers that would help vanquish males with smaller antlers. Such a scenario certainly is possible. However, Steve documented that across deer species there is a consistent mathematical relationship between the size of the deer and the size of its antlers; in particular, as a deer increases in body size, its antlers increase even more rapidly. It turned out that the Irish elk was just a very big deer, and a very big deer will have very, very big antlers as a consequence. Steve showed that the Irish elk has exactly the right sized antlers for a deer of its size. He further argued that the large size of the antlers did not likely derive due to selection on the antlers per se but rather due to natural selection for something else: large body size. Steve posited this because Irish elk evolved during the Ice Ages, an exceptionally cold time on this planet, and it is a well-known phenomenon, referred to as Bergmann's rule, that mammal body size tends to increase as average temperatures decrease.

Another reason the relationship between volatility and susceptibility to mass extinctions would have resonated with Steve was because his views on evolution, at least later in his career, placed substantial emphasis on the phenomenon of contingency (see chapter 4 for a more detailed discussion of Steve's views on contingency and how they changed throughout his career). Contingency is the notion that a single event in time, such as a mass extinction, can have a major effect on the subsequent course of evolution. Volatility means that mass extinctions are going to be especially important for explaining the types of organisms that persist on the planet over the long term. Steve further argued that the centrality of contingency means that the history of life will be largely unpredictable. This point was also recently amplified in the world of human affairs in the excellent 2021 Danish film *Riders of Justice*, starring brilliant actor Mads Mikkelsen. In a very important sense, we agree with Steve and *Riders of Justice* on this, yet we also partly diverge, and this is an issue that we will return to later in our essay.

VOLATILITY AND THE STOCK MARKET

Just as volatility can help us better understand the long-term dynamics of major clades in the history of life, it also can help us better understand the performance of different stocks in the stock market. Further, it helps illustrate some of the

commonalities between such disparate entities. (Our understanding of the relationship between volatility and the stock market was very much shaped by reading a 2011 paper by Malcom Baker, Brendan Bradley, and Jeffrey Wurgler, "Benchmarks as Limits to Arbitrage," as well as discussions Bruce had with Malcolm
Baker of the Harvard Business School.) It turns out that those in the investment
field have developed a term to describe the volatility of an individual stock, called
"beta," which represents the movement of the price of a stock relative to the rest of
the stock market. Stocks whose prices move more than average will have higher
betas, whereas those with less movement than average will have lower betas. It
also turns out that there was an approach to investing that was referred to as the
capital asset pricing model (CAPM), which advocated that the greater the
potential risk of the investment, the greater the potential reward. This
approach emphasized investing in risky stocks that tended to have large price
swings, that is, stocks that had high beta.

Baker, Bradley, and Wurgler considered the efficacy of this model and other
topics by looking at the performance of the top few thousand stocks in the market from 1968 to 2008. They split the stocks up into five equal categories or quintiles, from lowest to highest, and then examined how much money one dollar ($1)
invested in this group of stocks returned over this forty-year interval, adjusted for
inflation. What they found was surprising, at least to supporters of the CAPM.

They documented that investing $1 in stocks from the top 20 percent of beta in
1968 yielded a paltry sixty-four cents (adjusted for inflation) by 2008. Although
we are not well attuned financially, given that we are after all paleontologists, we
still realize that this comprises a very bad investment. By contrast, investing $1 in
stocks in the bottom 20 percent of the beta yielded $10.28, adjusted for inflation,
in 2008. This seems to us like a much better investment. And the results Baker and
colleagues found are exactly counter to the predictions of the CAPM. Indeed, anyone who used the CAPM investment strategy would have fared very poorly. By
contrast, an investment approach based on identifying low beta stocks would have
been much more propitious. Notably, some of today's most respected investors,
including the highly regarded Warren Buffet, never endorsed the CAPM and
instead focused on looking for stocks characterized by low beta, among other features. Significantly, this pattern shows much in common with what we have
described from the history of life. Just as in the long term it paid to have a low rate
of evolution, like worms and contra trilobites, it pays, literally and figuratively, to
have a low beta. In this case, we are treating performance on investment as a

measure of financial success, just like species diversity is a measure of a clade's evolutionary success. This is generally valid because a company's stock price relates in an important way to its financial health and overall level of success and performance. To take this argument further, companies whose stock price reaches $0 have reached a point of bankruptcy, akin to, but not necessarily synonymous with, extinction. In the history of life it's better to invest in a stock that embodies a worm's volatility rather than a trilobite's. Of course, we should say that we earnestly hope that no one uses any statements we are providing here to influence their investment strategy, the types of stocks they purchase, or anything else for that matter. . . . We are not professionals in investing, and we would hate for anything to go wrong with any part of your life savings. Instead, we urge you to leave things like investing to the professionals. We are simply struck by the parallels between clades and stocks, and we hope the reader feels similarly.

Another corollary aspect of clades of biological species and stocks is manifest when one examines Baker and colleagues' time series on investments in greater detail (figure 6.5). Just as the history of animal life is punctuated by five catastrophic mass extinctions involving species diversity falling dramatically, the history of the stock market is also punctuated by catastrophes, in this case, stock market crashes, when the value of stocks falls dramatically. Indeed, four times between 1968 and 2008 there were precipitous drops in the market, indicated by the "Xs" in figure 6.5. For instance, there is one shortly after Reagan was elected president in 1981, another toward the end of his second term in 1987, the crash after the 9/11 attacks in 2001, and the crash in 2007-8, precipitated by the demise of Lehman Brothers. We note solely for purposes of controversy and glibness that all four of these crashes occurred during sojourns by Republicans, and definitely not liberal ones at that, in the Oval Office. If conservatives tend to be richer and in favor of good times for the stock market, and therefore vote for Republican presidents accordingly, this may just further drive home the point that no matter what, people tend to vote in opposition to their own self-interests. That irony notwithstanding, recall that mass extinctions had particularly nasty implications for high volatility taxa. A very similar phenomenon emerges in the history of the stock market. In this case, it is the highest volatility stocks, those in the top quintile line in figure 6.5, that dip the most profoundly during market collapse. By contrast, the low volatility stocks in the bottom quintile of beta pretty much sail through those events unscathed and with only tiny dips in their investment valuation.

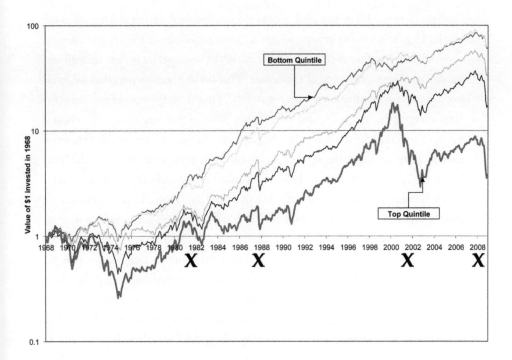

6.5 The returns for publicly traded stocks with >2 years of returns divided up into quintiles according to their beta. Top and bottom quintiles are our primary focus and are labeled, but intermediate quintiles are shown situated between these. The period covered is 1968 to 2008, and returns are based on a hypothetical $1, not adjusted for inflation, invested at the beginning of the interval. The *X*s represent stock market crashes, which are analogous to mass extinctions in the history of life. Figure modified from M. Baker, B. Bradley, and J. Wurgler, "Benchmarks as Limits to Arbitrage: Understanding the Low-Volatility Anomaly," *Financial Analysts Journal* 67 (2011): 40–54; thanks to M. Baker and B. Bradley and with permission from Taylor & Francis and the CFA Institute. All rights reserved.

The reason this is true is analogous to the situation with high beta species: high volatility stocks have a high probability of their price going up or down even during normal market regimes. Then, if for some reason the probability of a stock's price declining goes up dramatically across all stocks in the market, those high beta stocks will naturally be the ones that show the steepest price declines. Their prices may even decline so steeply that they reach a zero-price point, or bankruptcy, which is something analogous to extinction.

It is particularly striking that there are these profound similarities between species and stocks given that the two entities are quite different and diverge in fundamental ways. For instance, stocks do not speciate, except during splits of stocks. (Figure 6.3 shows genera, which are amalgamations of species, not species themselves.) Another difference between stocks and species is that sometimes bankruptcy is not the same as extinction. This is the phenomenon that we have become all too familiar with: "too big to fail." That is, in the case of certain very large companies, it was decided that if they were to truly go extinct the consequences would be far too severe for the economy. So, instead, our tax dollars were used to prop the company up and prevent it from going extinct. (Alas, the "too big to fail" philosophy did not hold in the case of the trilobites. . . .) Thus, sometimes extinct is not forever in the stock market. Last, there are certain underlying phenomena that favor the continual origination of high beta stocks, and that is because they are abetted by certain "irrational" investment strategies. For instance, people might know that there is very little chance that a stock will go up. However, if its price is low and it is showing a lot of variability in price, they might be willing to risk, even essentially waste, a dollar in the hope that very occasionally they get a big payback. This is akin to the lottery effect, where folks know there is an infinitesimal chance of winning the Lotto, but they will risk a few dollars on the very, very off chance that they might win a lot of money. Yet despite all these differences between stocks and species, the fact that they still show very similar patterns indicates that the same underlying processes could be operating. Steve made a related point that we will return to when we talk about his work on the decline of the .400 hitter and what it can tell us about the nature of evolution. Of course, there are many differences between evolution in the biological sphere and in the human sphere, see chapter 5. But what we have outlined here and regarding volatility is one significant similarity.

VOLATILITY AND STARS IN THE UNIVERSE

There are also very similar patterns relating to volatility in the rate of star formation and explosion (figure 6.6).

What figure 6.6 reveals is that the star formation rate and the star explosion rate have declined substantially through time. (This figure also reveals at least one salient difference between astrophysicists and paleontologists: astronomers read time with ancient events graphed on the right, and the modern graphed to the left, the opposite of paleontologists.) Twelve billion years ago, which is early

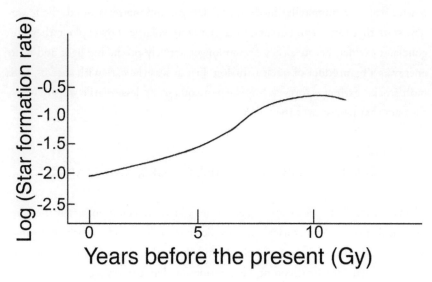

6.6 The star formation rate through time. From Bruce Lieberman and Adrian Melott, "Declining Volatility, a General Property of Disparate Systems: From Fossils, to Stocks, to the Stars," *Palaeontology* 56 (2013): 1297–1304, used with permission of John Wiley & Sons and the journal *Palaeontology*; thanks to A. Melott.

in the history of the universe and about how far astronomers can at present see back into the past, the star formation and explosion rate was ten times higher than it is today. (The y-axis is given in a logarithmic scale.)

What this means is that there are far fewer stars forming and exploding than there used to be. This is a good thing from our perspective and the perspective of life elsewhere in the universe, because massive stellar explosions, like supernovae and gamma ray bursts (GRBs), can have a seriously negative effect on planetary biotas anywhere in their vicinity. (See chapter 8 for a discussion of how these massive explosions have led to extinctions on this planet.) The universe, because of the decline in the rate of stellar explosions, has become a much more boring albeit a much safer place to live, and it is reasonable to posit that many life forms that may have evolved on planets early in the history of the universe may have unfortunately been blasted to smithereens by GRBs and supernovae that were too nearby said planets for comfort. Sometimes boring can be a good thing.

The reason that the star formation rate and the star explosion rate have declined through time relates to the fact that stars are historical entities, just like stocks and biological species. The formation of new stars removes gas and matter from the interstellar medium; the less gas and matter around, the fewer new stars that can form. Further, once the more volatile stars explode they are gone and extinct, because they are no longer actively producing light and heat energy as a by-product of nuclear fusion. Just as was the case with species, and most stocks, extinct is forever. It is the more quiescent, less volatile stars that are the ones that persist over the long term.

COMMONALITIES AMONG SPECIES, STOCKS, AND STARS

The key thing that species, stocks, and stars all have in common is that they are historical entities. As historical entities they have a birth and a death, and some duration in between. The more volatile an historical entity is, the more likely it will reach that zero, death point, be it extinction, bankruptcy, or explosion, and that is a point of no return. It turns out these entities are especially at risk during mass extinctions, financial crises, and the early history of the universe. So it seems to be nearly universally true that the tortoise wins the race.

What we have discussed here seems to represent a general pattern that holds across seemingly very disparate areas, but of course there are many such patterns. One of the most famous goes back to a phenomenon identified by the great Italian scientist Galileo. He recognized that the relationship between surface area and volume has important implications for the structure of living things because as organisms get bigger, their volume or weight increases more rapidly than their surface area. This is restating a simple mathematical relationship because volumes increase as a power of 3 while areas increase as a power of 2. Surface areas are important to living things because that's how they obtain food and oxygen; volumes are important because those are the interior cells to which the food and oxygen need to be distributed and they comprise weight. Further, the surface area of a bone determines how much weight, or volume, that bone can support. Thus, significant features relating to the appearance of organisms, their shapes, and constraints on their size are all determined by this mathematical relationship. Steve, in a stroke of what can only be described as sheer genius, recognized the significance of this relationship, and extended it further and to something outside of biology, specifically, the architecture of Gothic cathedrals. This was documented in his essay "Size and Shape" in the classic *Ever Since Darwin*. It was fascinating and wonderful that Steve could document the same basic relationship for things as disparate as buildings and organisms. That human life and human buildings are beholden to and governed by the same basic mathematical rules suggests that, to borrow one of the phrases Steve used in a slightly different context, we are "of it, and not above it," that is, the human essence is fundamentally natural and not supernatural.

Another area where Steve was able to take a principle or pattern from one area and apply it to another involved his love for the game of baseball. He specifically considered the prevalence of .400 batting averages in major league baseball. Batting .400 over the course of an entire season is a truly noteworthy and impressive feat that indicates a very high degree of skill. It was an occasional but regular occurrence back in the nineteenth and early twentieth centuries. Today, however, it's something much rarer than witnessing a steam locomotive pulling freight down a mainline track. In fact, the last person to bat over .400 was Ted Williams, who achieved this feat in 1946.

Steve pondered why this was the case, first in a 1986 essay in *Discover* magazine, then more fully in his presidential address to the *Paleontological Society* published in the *Journal of Paleontology* in 1988, and then in his 1996 book *Full House*. One of the traditional explanations for this pattern was that baseball players used to be better in the august, mythic past. However, there are a variety of reasons, including modern training regimens, diet, the recognition that athletes have improved in all other sports, and so forth, to think this is not a valid explanation. Instead, Steve argued, the real reason .400 hitters have vanished is a corollary to the fact that .160 hitters have vanished as well: the "average" batting average today is very similar to what it was in the past, .260. However, the overall variance of major league batting averages has declined.

Steve saw the contraction of variance around a constant mean as a common property of many different systems. He held that systems were often characterized by early experimentation that led to a plethora of variation, yet the outliers in the spread of variation were later pruned, leaving far fewer distinct types as things became canalized and the system stabilized. For instance, he saw this as a quintessential aspect of the Cambrian radiation and its aftermath, when there was an early burst of many body types that later became pruned to a more circumscribed cadre of some twenty phyla, discussed in detail in his book *Wonderful Life*, and the history of the automobile, where early on there were many different types of power sources that were used for engines, but eventually only one came to dominate, the gasoline-powered. This appealed to him because there was no need to invoke any special adaptive or particular explanation for any single case. Instead, the explanation was broader and more universal, holding for many cases. When it came to the many phyla in the Cambrian that subsequently died off, it is true that each died for a reason. However, the general expectation should be, according to Steve's argument, that far fewer should be left standing than initially evolved. Thus, we think Steve would have liked our argument about volatility, though sadly he passed away before we got the chance to ask him.

WHAT IS PREDICTABILITY, AND WHAT DOES IT HAVE TO DO WITH STEVE'S POLITICS?

One of the big questions in evolutionary biology is to what extent is evolution predictable. This is part of a broader topic which gets to the very nature of science and the possible differences between historical and experimental approaches to science. For instance, chemistry tells us that if one adds water to a strong acid there will be an explosion. This is the reason one needs to follow that chemical mnemonic, "do as you oughta, add acid to water." Physics tells us that as an object falls to Earth it will accelerate (though wind resistance can eventually cause that acceleration to cease). This is the reason you want to avoid falling from a great height. Notwithstanding that, it was the great physicist Niels Bohr who remarked that it is "very hard to make an accurate prediction, especially about the future." But basic Newtonian physics means that the behavior of certain things, in a physical sense, is generally predictable.

What does Darwinian evolution entail about predictability when it comes to the history of life? Well, when one enters any arena involving history and its forward casting one might be quick to suspect that all bets are off. Yet to the extent that one accepts George Santayana's clever admonition "Those who don't learn from history are doomed to repeat it," it suggests that humanity could indeed learn from the past to augury the future. Regrettably, humanity's track record of learning from the past is poor at best, and indeed, no less a personage than George Bernard Shaw supposedly remarked that "the only lesson that can be learned from history is that no one learns from history." Does this mean that according to Santayana we are doomed, or instead are we merely trapped in a tautological purgatory? Alas, neither Shaw nor Santayana are available to resolve this conundrum.

In a general sense, when it comes to perhaps the two greatest theories of the nineteenth and early twentieth centuries regarding the predictability of the development of human minds, civilizations, and economies, Freudianism and Marxism, respectively, both have failed miserably. It is perhaps intriguing and ironic, given the tenor of Marxism and its supposition that human history follows a definite path, that Steve was a fan. Recall that at numerous times throughout his career he endorsed, though not always consistently, the phenomenon of contingency and how it adversely impacted our ability to predict the history of life. This is a topic that we already presented on in chapter 4. Why should he endorse Marxism and its associated predictability in the purview of human history, whereas in the equally complex domain of the history of life he eschewed it? People can be a bundle of contradictions, and Steve was no exception. Maybe when you're a Harvard professor who gets hired in the late 1960s being a Marxist is just something that's expected, like a large salary and a posh place in the lovely city jokingly referred to as the People's Republic of Cambridge. This is only part of the explanation, of course. Steve was often fond of stating that he learned Marxism at his daddy's knee. Thus, family played a big role in his politics, as it does with anyone.

Another aspect of Steve's Marxism and these sorts of contradictions inherent in any person can be indicated by examining fellow Harvard professor Richard Lewontin, who was legitimately brilliant and Steve's close role model. Lewontin espoused his viewpoints in various places including the book *The Dialectical Biologist*, written with Richard Levins. Lewontin was undoubtedly a committed Marxist in many respects, yet there were limits to his Marxist stolidity. For instance, a 2021 epistle to the Harvard alumni magazine revealed that at least one time Lewontin didn't respond to a letter sent by a student because he had never seen it. The reason, it was sent via campus mail, and thus sans stamp; Lewontin told the student "he did not read campus mail: if the sender was not willing to pay for a postage stamp on the envelope, it was not worth reading." Sometimes capitalism can rear its ugly head and thwart communication, even between advisor and former student. Maybe in the case of Steve the apple doesn't fall that far from the tree. But Steve was very fond of quoting that "a foolish consistency is the hobgoblin of little minds," which derives from Ralph Waldo Emerson, who went on to add, "adored by little statesmen and philosophers and divines." We think Emerson, and by extension Gould, nailed it.

From our seats of assessment, not of judgment, a key thing is to try to be as caring and kind as possible and leave the world a better place. And Steve *was* often kind and did *care* about the fate of humanity. He truly left the world a better place, and he did try to help those individuals that might have been at either an economic or other disadvantage relative to himself. For instance, he and his then-wife Rhonda Roland Shearer handed out blankets and foodstuffs to those participating in the post-9/11 search and rescue efforts. Their contributions were significant enough that the New York City Firefighters Pipe and Drum Band played at Steve's memorial ceremony held at New York University.

Yes, when Steve moved to New York City's SoHo, once the neighborhood of artists and quirky types mercilessly parodied in Martin Scorsese's flick *After Hours*, that later became thoroughly gentrified and deluged by rich and/or famous tenants and hordes of much less rich and/or famous tourists, he came to inhabit a valuable piece of high-ceilinged, hard oak-floored property that had been expensively remodeled. Its tastefully appointed nature seemed at odds with the lifestyle of a self-proclaimed Marxist. Yet at the same time, Steve also diverted royalties from his books to support his graduate students, something that is otherwise unheard of, at least to us. This was to thwart criticisms that he was not a grant-getter of any accomplished extent. Also, he told Niles in Dijon in 1982, he was worried that he would be taken as just writing for the money. Niles reminded Steve of his recent car purchase, which, if he had been in it for the money, would have been a Jaguar or something of similar vintage, while instead Steve had bought a Dodge Dart.

Better that Steve should have been a rich, caring Marxist than a rich, uncaring Social Darwinian. In our view, if you earn the cash, you should be allowed to spend it, and at least Steve never forgot his middle-class roots and that it was partly a matter of luck or contingency that explained why he could own such a loft while others couldn't. And he did very much take to heart advice from Bruce, a home-grown, early, now erstwhile resident of SoHo, who told him to be sure to shop at Vesuvio's bakery and Joe's Dairy (the latter entirely defunct, the former defunct, though thankfully they have preserved the window, awning, and outside appearance). Indeed, upon hearing this, Steve theatrically removed a tiny, black spiral bound notebook and pen from his inside jacket pocket and jotted down those tidbits offered to give him a more authentic, neighborhood experience. Given that this conversation occurred when Bruce was still comparatively a wee bairn without a tenure-track academic job, he could have been amazed that Steve wrote down anything he had to say. That he wasn't can be ascribed to the fact that when it came to this sort of thing, Gould lived up to his Marxist precepts and was definitely antihierarchical.

GIVEN CONTINGENCY, IS EVOLUTION PREDICTABLE?

Transitioning from the evolution of societies to the evolution of life, that foil of Steve's, Richard Dawkins, has argued that much of the history of life is predictable. However, he has been shown to be wrong by numerous sources, including the kids from *South Park*. We will not consider him further here (though see chapters 5 and 12). By contrast, as already mentioned, Steve argued in numerous places that contingency rules the history of life and thus predictability is out the window. We largely agree with him. However, sometimes the very existence of contingency can lead to predictability. For instance, fossils, stocks, and stars are all historical entities, so their behavior is contingent, yet that is also precisely why we can predict their behavior, at least given some information on their volatility. Contingency leading to predictability? We'd be lying if we told you we knew anyone who predicted that.

When it comes to predictability, we'd further argue that even though there is not an accurate, scientific way of guessing the precise species composition of the biosphere fifty or even five million years in the future, because we can't predict the pathway of evolution, there are certain predictions that evolutionary biologists and paleontologists can make. For instance, the predominance of the allopatric mode of speciation, a key evolutionary principle, allows us to predict that a visit to the beautiful Azorean island chain will afford the naturalist an opportunity to observe several new species. This is because in general islands are a place where one finds new species, since they are geographically isolated and thus centers of allopatric speciation. We might even predict that at least one of those species would be a bird, related to another bird found on the closest large body of land, because some birds fly great distances and thus could have reached an island chain from the mainland and subsequently become isolated. And if we did make such a conjecture, we'd be correct.

We might also predict that we would find a unique flowering plant, since certain plants can have extensive dispersal capabilities associated with the part of their life history they spend as seeds. Again, we would be right to make such a prediction. However, we would be deluding ourselves and others if we claimed that we knew in any precise detail what these distinctive species would look like, at least until we actually saw them.

A key part of Darwin's early ideas on evolution, documented in the *Voyage of the Beagle* and the *Darwinian Notebooks*, was his recognition of this principle of allopatric speciation. For instance, there are Darwin's observations on the two species of large flightless birds, rheas, known from South America, which had adjoining but mutually exclusive geographic distributions that met on either side of the Rio Negro in Argentina. There is also the tragic example of the Falkland island wolf (sometimes called a fox; Darwin called it a "wolf-like fox"), a now extinct, quondam inhabitant of this bleak island chain—tragic because the beast's tameness and naiveté to the pernicious danger humans represented made it exceptionally easy to approach and kill, extinct only decades after humans arrived. Darwin reputedly collected several specimens with a geological hammer by bonking them on the head when they were in close range.

There are other reasons why the general pathway that evolution follows may be constrained or directed along certain lines. Although this doesn't quite mean it's predictable, it means we've got more than just the foggiest of notions about what to expect. One reason is related to the phenomenon of convergence, whereby organisms with different evolutionary histories come to resemble one another, to a greater or lesser extent. Convergence occurs because there are only so many ways for an organism to make a proverbial living. For instance, if an animal is going to spend a good portion of its life moving around off the ground it's got to fly, and therefore it's got to have wings. Another biological reality that narrows down the number of evolutionary pathways is the existence of constraints. Evolutionary constraints entail that there are only certain types of forms that can be produced. Consider flying animals and wolves. Seemingly the wolf lineage can never join the flying club because they can't evolve wings, given that their forelimbs, and indeed those of the taxonomic family they belong to, the Canidae, are decidedly unwinglike and instead well constructed for running at a good pace over long distances. When these constraints are confronted by ecological challenges and opportunities, there are certain forms that evolve again and again. Steve championed this viewpoint, citing what he termed Galton's polyhedron, and suggested that evolutionary change is more like casting a die than rolling a perfectly spherical marble. In particular, when a die is cast it must land on a specific side, and there are only so many sides—forms that an organism can come to represent—unlike the effectively infinite number of points that exist on a smooth sphere. We are rebranding Galton's polyhedron as Gould's polyhedron in chapter 9, which discusses the repeated evolution from a wolf ancestor of a form intermediate in size and appearance between coyotes and wolves.

Another prediction we can make is that a major episode of climate change or geological change will have a profound impact on the history of life. Even in our day-to-day existence we can feel climatic extremes and observe their collateral nefarious side effects (heat that leads to drought, that leads to large scale fires, etc.). As paleontologists, we may see a neighboring mountain that we need to reach for fossil collecting purposes, but behind it stretches a gathering storm that strikes trepidation, not least because occupying high elevations on a barren surface of rock four hundred miles from civilization and the nearest accredited medical doctor while holding a geological hammer is an inordinately bad time to try to recreate Benjamin Franklin's early experiments on the nature of electricity and lightning. To bring the point home, those very same rocks may happen to preserve evidence of environmental changes that lasted much longer and were inestimably more profound than a single, localized storm; those antediluvian changes in turn led to a surfeit of extinctions changing and resetting the course of animal evolution for the next five hundred million years. Contingency is wonderful, and so is making it home from the remote field site in the Arctic Circle referred to in the previous few sentences. What was Steve's primary field site? Bermuda. We assert that he chose wisely there.

CHAPTER 7

Paleo Personas

Musings on a Soviet Cephalopod, Norman Newell, and Mass Extinctions

Being an academic has several perks. One of the best is the opportunity to work with talented students. Serving as a museum curator also has its rewards. Bobby Bacala of *The Sopranos*, played by Steve Schirripa, once remarked that after manicurist, museum curator was the profession most anathema to successful gangsters. Neither of us has ever gotten a manicure, but opening up a drawer packed with specimens of a modern species (figure 7.1) or ancient ones (figure 7.2) can be awe-inspiring, much like staring up at the night sky and glimpsing a distant nebula so far away that it took light seventy million years to travel from those stars to your retina. We museum curators know that often a good way to find something astonishing, even something "new," is to browse through old, neglected drawers in natural history museum collections. An excellent example of this involves a specimen of the trilobite *Bouleia dagincourti* (figure 7.3) found by Niles in just such a neglected drawer in the American Museum of Natural History (AMNH). It had been given to Norman Newell, about whom we will hear much more in this chapter, by archaeologist Junius Bird.

Originally, *Bouleia* was thought to be very closely related to *Phacops*, a genus of trilobites discussed in more detail in chapter 2. However, analysis of the morphology by Niles demonstrated that the arrangement of muscle scars on the heads of these two types of trilobites was quite different, revealing that they were much more distantly related; indeed, they belonged to entirely distinct families. This would be equivalent to showing that what were once thought to be two species of cats in fact represented just one cat and a member of the weasel family.

The science of paleontology intrigues so many because it is a science that provides a window on the past. Bruce obtained a glimpse through such a window via the work of one of his graduate students, Kayla Kolis, who received her master's

7.1 House sparrows in the collections of the Division of Ornithology, Biodiversity Institute, University of Kansas (KU). Photo by Mark Robbins (KU).

7.2 Cretaceous fossil echinoids (sea urchins) in the collections of the KUMIP. Photo by Bruce S. Lieberman.

degree in 2017. As part of her thesis she was studying patterns of evolution in cephalopods that lived roughly 310 to 270 million years ago, during what geologists refer to as the Pennsylvanian and Permian periods, in seaways that covered parts of North America. These animals are relatives of the modern squid and nautilus.

Most of the species of fossil cephalopods Kayla studied, such as those depicted in figure 7.4, were unique to that part of North America, as might be expected. However, one of the specimens in the collection of fossil invertebrates at the University of Kansas Biodiversity Institute (KUMIP), a fragment of a nautiloid that stretched out like a crooked finger (figure 7.5), had a more surprising pedigree. A glimpse at the labels (figure 7.6) associated with the specimen indicated she had made a startling discovery.

7.3 Image of a latex cast and internal mold of a specimen of *Bouleia dagincourti* from the Devonian of Bolivia, in the collections of the American Museum of Natural History. Picture from N. Eldredge, "Morphology and Relationships of *Bouleia* Kozlowski, 1923 (Trilobita, Calmoniidae)," *Journal of Paleontology* 46 (1972): 140-51. Image used with permission Society for Sedimentary Geology (SEPM) and courtesy of Kathleen Huber.

7.4 A complete specimen (from the KUMIP) of *Metacoceras jacksonense*, a nautiloid from the Pennsylvanian period of Kansas. Photo by Bruce S. Lieberman.

This specimen came from a country that no longer exists today, the Soviet Union, albeit one of its descendants, Russia, is still very much in the news. Moreover, it was collected in 1937 by a highly accomplished paleontologist, Norman Newell, who spent many decades working at the American Museum of Natural History, where he served as the PhD advisor to Niles and was thus the academic grandfather to Bruce. He cut an impressive figure throughout his many years at the AMNH. How he ended up in the Soviet Union in 1937, during the height of the purges, was a big surprise, but his former thesis advisor and later boss at the University of Kansas, Raymond C. Moore, sent him in his place to the Seventeenth International Geological Conference in Moscow that July. More aspects of his voyage may lie buried in the archives of the American Museum of Natural History's library, which at present is under construction so that we could not obtain access to elucidate this mystery further.

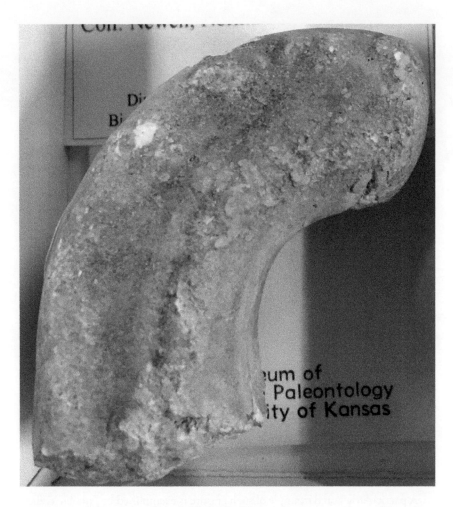

7.5 An incomplete specimen (from the KUMIP) of *Metacoceras*, preserving a part of the coiling shell of a Pennsylvanian nautiloid from the former Soviet Union. Photo by Bruce S. Lieberman.

A sordid and disturbing aspect of this time in the Soviet Union was the penchant for people to simply disappear at the behest of Stalin and the state security system, which was embodied by the secret police, or NKVD. People were made to not just figuratively but literally disappear from the record, all traces erased, even more thoroughly than certain clades of long extinct organisms were eliminated during ancient cataclysms. What it would have been like to visit the Soviet Union at this time could have been equal parts horrifying and fascinating, and we only wish we had asked Newell about it before he passed away.

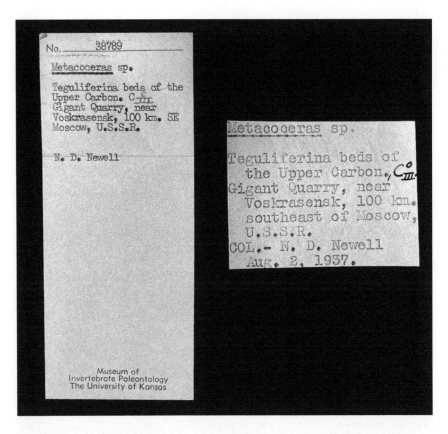

7.6 The labels associated with the incomplete KUMIP *Metacoceras* specimen from the Soviet Union. Photo by Natalia López Carranza (KU).

A particularly noted example of such erasure is exemplified by Nikolai Yezhov, who was the head of the NKVD from 1936 to 1938. There is a famous picture of Yezhov, seemingly strolling happily by Stalin's side on the banks of the Moscow–Volga Canal. Soon after, Yezhov was purged and erased from all Soviet documents after he became the object of Stalin's enmity. A later version of the photo after Yezhov's execution in 1940 shows the same scene, but with Yezhov removed. At Stalin's side there was now only empty space. This was long before Photoshop, so it would have been no trifle to arrange, but Yezhov had been artfully and manually eliminated from history.

Not that Yezhov deserves any sympathy, for he himself was responsible for erasing and killing hundreds of thousands of innocent people, either by having them murdered outright or by having them sent to the gulags, a fate worse than death.

But the simple concrete evidence preserved in the case of the two pictures, much more surreptitiously done in most other instances involving less well documented persons, illustrates the notion of how people could be made to simply disappear and is truly horrifying.

Such disappearances were not unknown among Soviet geologists. One of the scientists Newell was likely very much hoping to see at the 1937 International Geological Conference was Dmitrii Ivanovich Mushketov, a towering figure in Russian and indeed international geology. However, a month before the conference Mushketov was scooped up by the NKVD and made to disappear under trumped-up charges of "organizing a counter-revolutionary fascist terrorist group in 1930 and committing acts of sabotage." Mushketov was shot eight months later. It may well have been a jealous geologist who viewed Mushketov's talents as a threat and envisioned a sordid opportunity for advancement who dropped the proverbial dime on him. (We learned about this in a fascinating article on Mushketov by Yu Solov'ev of the Russian Academy of Sciences in the journal *Stratigraphy and Geological Correlation*.) Maybe sometimes geologists, and even museum curators for that matter, are not so far from removed from gangsters as everyone would like to believe? Usually, however, when academics take steps in such a direction it doesn't involve physical violence: instead, they metaphorically erase scientists who challenge them, outshine them, or otherwise cause them discomfort. They tend to do this by trying to block individuals from their and thereby others' minds, often by ignoring that person's work entirely. This could include failing to properly cite or acknowledge their work, a topic we discuss in greater detail in chapter 11. In all events, Mushketov was exonerated in 1956.

Now we turn from tales of individuals being erased to ancient cataclysms that vanquished long-extinct clades. And here the story and place of paleontological luminary Norman Dennis Newell (1909–2005) become much clearer.

He was raised in the great state of Kansas. His mother drove him to the University of Kansas (KU) in Lawrence—if Niles's memory serves, Norman said it was in a horse-and-buggy—in 1925. There he studied geology, receiving his bachelor's and master's degrees under the tutelage of paleontologist Raymond C. Moore, state geologist and KU geology chairman. He was also a skilled clarinetist; later in life Norman would occasionally get out his old clarinet to play a few "licks," which further endeared him to our hearts and those of others who were his academic children and grandchildren and also shared a love of jazz.

Norman became entranced with the Upper Paleozoic fossil riches of Kansas and was drawn especially to clams (bivalved mollusks). He wrote his doctoral dissertation at Yale under the general tutelage of Carl O. Dunbar, who is discussed in chapter 13. The famed Charles Schuchert, now retired, was still hanging around, and Norman once said to Niles that the old boy had become hard of hearing, so Norman got to hear all his words through the wall of his next-door office as he shouted out what must have been generally fascinating and engaging stuff into the telephone.

Norman worked in Kansas for the state geological survey and later as a faculty member at KU in the 1930s. (He moved to the University of Wisconsin in 1937.) His 1937 publication on Upper Paleozoic Pectinacea (scallops and their scallop-like forebears) remains a classic of how one looks at evolution through the lens of a paleontologist.

Norman spent three years during World War II mapping the geology and collecting the fossils around Lake Titicaca for the Peruvian government. His specific charge was to locate exploitable oil deposits. As far as we know, none were to be found, but when Norman accepted a job offer from George Gaylord Simpson to become an invertebrate paleontologist at the American Museum of Natural History in 1945, he brought a choice collection of his Peruvian fossils, including trilobites. Years later, in the late 1960s, Niles found them just sitting there, waiting to be studied so their secrets could be revealed.

In the 1950s Newell, teaming up with the Smithsonian's brachiopod specialist G. Arthur ("Gus") Cooper, pursued a massive, innovative study of the silicified Permian fossils of the aptly named Glass Mountains of Texas. Joining them was an array of colleagues, graduate students (mostly Norman's Columbia kids, including Roger L. Batten, Donald Boyd, Robert M. Finks, Ellis Yochelson, and Keith Rigby), and technicians. That endeavor led directly to Norman's perhaps even more famous team effort to understand the carbonate ecology and physical environmental regimes of the Bahamian platform, a pioneering study of modern marine ecology sparked by his passion to understand all the better those ancient Permian environments.

But probably above all else was the fact that by the 1960s Norman D. Newell had become almost the sole and certainly the most important figure to proclaim the reality of mass extinctions. He called them "crises in the history of life." As later emerged from a ton of data and careful analysis, the end of the Permian saw the greatest of all mass extinctions to engulf life on Earth since the emergence of complex, multicellular life more than a half-billion years ago. Norman had also become an expert on gaps in the stratigraphic record of time, what geologists term unconformities, paraconformities, and the like. He knew he had to battle the traditional litany of complaints asserting that what the fossil record seemed to show about the history of life could be explained away because the record is riddled with missing time—ergo missing sediments and fossils, missing data. Newell had to convince himself first, and then the rest of us, that despite missing time the signals sent by the fossil record are real records of actual events and not just the artifact of poor temporal preservation.

Niles, by then a very junior curatorial member of Norman's small invertebrate paleontological department at the AMNH, once had a long tête-à-tête in Norman's office. That would have been sometime in the early to mid-1970s. The conversation was cordial but also quite intense, and even (on at least two points) nearly heated. Niles was expounding on the virtues of cladistics, the then-newfangled approach to phylogenetic analysis that had been sweeping the AMNH in the late 1960s, and was beginning to seep through its thick stone walls to the American hinterlands. Norman said that it was all very interesting, but especially since the young Turks espousing the approach were willing *not* to include temporal (stratigraphic) information in their data sets they were needlessly handcuffing their problems to make them mimic the approaches of systematics to purely contemporary, living taxa—for which phylogenetic systematics was originally devised by the German entomologist Willi Hennig.

Norman was defending the importance of time in thinking about evolution. He wasn't wrong of course: Niles, despite having produced a cladogram hypothesizing a pattern of relationships among the various subtaxa of the Middle Devonian *Phacops rana* trilobite "species" (as was then thought), had also at the same time produced charts of stability and change in time and space of these fossils, the empirical basis of the notion of "punctuated equilibria" discussed in some detail in chapter 2.

Norman was also a great teacher, and for a while in the mid- to late 1960s he convened with Batten a weekly seminar focused on a different group (e.g., mollusks, arthropods, etc.) each semester. Graduate students of two age classes were in regular attendance: in Niles's younger entering-group of 1965 there were also Gene Gaffney, John Boylan, Bob Hunt, and others. But there were those as well who had entered the program in 1963 and taken Niles under their wing while he was still a college junior, including two whom he has long since come to regard as

7.7 From left to right: Stephen Jay Gould, Norman Newell, and Niles Eldredge at Norman's ninetieth birthday party in his former office/lab at the American Museum of Natural History in February 1999. Image courtesy of Gillian Newell, used with permission.

his academic older brothers: H. B. Rollins and Stephen Jay Gould, a.k.a. Bud and Steve (figure 7.7).

Steve was also a pivotal figure in Bruce's career and intellectual development, as we have already described. And Bud graciously served as the external committee member on Bruce's PhD thesis and provided lots of valuable insight.

Niles will never forget one day, back in the student's research room at the American Museum, after one of these seminars, Steve's saying "I swear that man [i.e., Newell] will go to his grave denying natural selection." It's somewhat ironic to think that this came as criticism from Steve, given that decades later he was castigated by some scientists for doing the very same thing; maybe aspects of his perceived intellectual heresies on this subject were partly inherited. Newell had probably been waxing eloquent on his love for the stratigraphic vision of the history of life. And maybe he was on a tear about extinctions as well. Niles remembers wishing we could talk about something positive (evolution) rather than negative (extinction) during those days. Little did Steve or Niles know then that important parts of their careers would be spent developing and promoting the idea that nothing much happens in evolution unless and until extinction, of varying intensities, shakes up an otherwise stable system.

Bruce's encounters with Norman Newell were much less lofty and far more indirect. Certainly he was impressed as a twenty-something graduate student to see him coming into work several times a week long after retirement and well into his late eighties. Further, the image of Newell taking a protracted, measured perambulation down one of the longest hallways in New York City, on the museum's fifth floor, carrying a small box of Permian bivalves to be photographed brings him back to fond memories of life in graduate school. One might step in and out of one's office several times to find him still moving slowly but steadily toward his destination. To have long-dead clams imbue one with such a sense of purpose is a testament to the triumph of science and the human spirit.

NEWELL'S LEGACY: MASS EXTINCTIONS

Extinctions are a central part of paleontology's fundamental connection to evolutionary biology. Indeed, starting with the work—and arguments between—the two great French naturalists Georges Cuvier and Jean-Baptiste Lamarck at the Jardin des Plantes in Paris in the early 1800s, extinction was perhaps the first great theoretical contribution paleontology made to evolution. However, before Newell's time, the notion that *mass* extinctions could occur was largely anathema to biologists. Indeed, Charles Darwin, in a last-minute comment scrawled into the "Fair Copy"—the clean handwritten copy made by an amanuensis and readied for the printer, though in this case never delivered—in his 1844 "Essay," wrote, "Better begin with this: If species, really, after catastrophes, created in showers over world, my theory false." Darwin was evidently jolted by the appearance in 1842 of a monograph by Alcide d'Orbigny, a student of Cuvier and rival naturalist who preceded Darwin by a few years in South America, on Jurassic ammonites. D'Orbigny introduced the concept of "stages" (étages)—bodies of rock recording intervals of time, demarcated and defined by the abrupt, near-simultaneous extinction and subsequent replacement of species. He also thought that his stages were worldwide in extent (*monde entier*, Darwin's "over world"). We now see that "stages" are not global: there have been fewer than ten truly global mass extinctions, including the human-caused one that we are now living through and hopefully surviving. Yet large-scale regional "stages" remain a key element of biostratigraphy. They are the basic units recording the "turnovers," births and deaths of species in environmental event-driven episodes. Norman Newell was the first scientist to create a database to study the phenomenon of mass extinction in detail.

We cannot agree with Darwin that regional and global mass extinctions prove his theory of evolution through natural selection wrong. But they do represent an important challenge to aspects of Darwin's views that posited that evolution is largely a slow and gradual process, with speciation and extinction events stretched out through time and competitive factors playing a primary role in each of these phenomena. Newell played a central role in developing the narrative that came to convince scientists of this challenge. He notably considered the effects of mass extinctions on reef-forming organisms, a topic very relevant given that present-day coral reefs are key sources of marine biodiversity and are also critically in peril. When we consider the paleontological context of reefs, at any one time over the last five hundred or so million years different phyla such as sponges, brachiopods, mollusks, or cnidarians comprised the dominant reef-forming organisms. It always seemed that eventually a mass extinction crisis came along to knock one phylum off the top of the heap; eventually, tens of millions of years later a different phylum would rise to the top.

Norman Newell, along with his wife, Valerie, would occasionally throw large parties for students and faculty involved with the Columbia/American Museum of Natural History program in paleontology. The Newells had a magnificent pool table in their basement, which was an important attraction at such events. In the late 1960s the Newells had attended yet another International Geological Congress in the Soviet Union, and Norman decided to regale his attendees at one such occasion with what turned out to be, shall we say, an exhaustive slide show of their trip. Exhausting is perhaps the better word. As the show wore interminably on, at last one of our graduate student numbers, the redoubtable Eugene Spencer Gaffney (soon to become the AMNH's fossil reptile curator, replacing the retiring Edwin H. Colbert), plucked a white handkerchief from his pocket and began waving it in the universal sign of surrender. At last Norman got the hint and concluded the program.

But Norman's show is far from over, so vital was he in so many ways in changing our understanding of how the stratigraphic history of life, riddled as it is with local, regional, and truly global mass extinction and evolutionary recovery events, connects with our grasp of the evolutionary process itself. Steve throughout his career helped keep that show running by keeping the focus on the paramount importance of mass extinction for understanding the history of life. Whether it was his foray into deducing the significance of the Burgess Shale for understanding pivotal themes in the history of life or his long-term flirtation with time's arrow/time's cycle and contingency recounted in chapter 4, an important aspect was passing on Norman Newell's legacy.

CHAPTER 8

Stardust Memories

Reading Evolution and Extinction in the Stars

One of the fundamental lessons of the fossil record is that extinction happens. Quite frequently (every five to ten million years, more or less regularly) this plays out in dramatic pulses of extinction, part of what Elisabeth Vrba of Yale University termed "turnover pulses," involving a good portion of species in local regions disappearing in a geological instant.[1] Elisabeth made many fundamental contributions to evolution and paleontology, and the turnover pulse hypothesis is certainly one of these. The way we think about the turnover pulse is as a scaled-up, multispecies version of punctuated equilibria, a.k.a. punk eek; Stephen Jay Gould mostly viewed it that way in his 2002 book *The Structure of Evolutionary Theory*. If you've made it this far in the book you know the drill about punk eek, but for those who skipped earlier chapters, punk eek is about the history of individual species and its relevance to evolution. For most of that history, which may comprise millions of years, they change little and display stasis. When change does happen it is associated with speciation, which occurs when a population of the species become geographically isolated from the rest of the species. Given enough time and consistent separation, that new population can diverge and form a new species; the formation of the new species is geologically rapid but staid ecologically, taking thousands to tens of thousands of years.

One key aspect of the brilliance of Elisabeth and her turnover pulse hypothesis is that she recognized that species rarely if ever live on their own. Instead, they occur in groups or regional biotas where many types of species are commingled: trilobites with brachiopods, trees with monkeys, bumblebees with flowers, and so forth. Thus groups of many species living together might be expected to experience stasis in concert. Further, the types of circumstances that might cause populations to become geographically isolated in one group, let's say trilobites, will also cause the associated populations of brachiopods to become geographically isolated as well. So we might expect that speciation events in different groups will be concentrated in time: Elisabeth's pulses.

THE LION KING MEETS BEAVIS AND BUTTHEAD

Elisabeth further concentrated on the specific types of factors that might lead populations to become geographically isolated from one another and/or the rest of the species. She recognized that changes in the environment, including climate change, would be critical. Elisabeth is an expert on many things, but the organisms she studies are fossil and modern tropical mammals of Africa, especially antelopes and their relatives. Data she collected on these indicate that over the last 2.5 million years the distributions of African tropical rain forests, and the mammals (and other creatures) they contained, were very much influenced by the waxing and waning of the planet between warm and cold intervals. The last three million years covers the interval colloquially referred to as the "Ice Ages." In terms of average temperature, this three-million-year period is the coldest it's been over the last five hundred million years. During the Ice Ages, traditionally consigned to what is referred to as the Pleistocene epoch of Earth history, ice sheets would periodically extend out from the north and south poles and mountainous regions as temperatures fell; in North America they once reached as far south as modern-day Kansas. By contrast, during the warmer periods within this 2.5-million-year interval, average temperatures were actually warmer than today, so much warmer that 130,000 years ago hippos were wallowing in the Thames and its tributaries, whereas today, of course, hippos are restricted to aquatic habitats in places in Africa with much balmier climes; during these warmer times the ice sheets retreated back toward the poles and higher elevations.

Just as these ice sheets waxed and waned, they even affected the climate in faraway places. Colder intervals also became drier, and consequently tropical rainforests contracted, while during the warmer intervals conditions usually became wetter, leading to rainforest expansion. Concomitantly, the mammals and other creatures that lived and depended on these forests had their geographic ranges ebb (when it was cold) and flow (when it was warm). Yet ebbing doesn't precisely describe how the ranges were affected. Elisabeth instead noted the forests became packed into smaller refugia separated from other such forest refugia. The change in the geographic range of a large tropical rain forest wouldn't best be depicted by a great circle that simply shrunk, but rather by a circle that transformed into several tinier circles scattered within the confines of the original great circle. Voilà: numerous tailor-made opportunities for geographic isolation among small populations of forest mammals.

This isolation of smaller, more restricted populations will naturally lead to an uptick in extinction and speciation in such creatures (and other co-occurring ones). The amount of extinction accelerates partly because smaller population size and geographic range makes species more likely to go extinct, and partly because species become trapped in a narrow place with no possibility of migrating as conditions change; the increase in speciation is just related to what we might call the law of geographic speciation, which holds that new species are prone to arise in isolation. Later, as climatic conditions ameliorated, and the forests expanded anew, those new species would spread out as well. This is an idea that has relevance not just to any old fossil mammal but also to the origins of our own ancestral hominid lineage, including the genus we belong to, *Homo*. It may have been a turnover pulse driven by climate change that triggered key divergence events in hominid evolution around 2.6 million years ago, associated with the appearance of the genus *Homo*.

Turnover pulse indicates that the pacemaker of evolution will be changes in the physical or abiotic environment. By contrast, the traditional emphasis in evolutionary theory has been on the primacy of competition and biotic interactions among species and the individuals that most foment evolution. This idea is certainly reflected in Darwin's *Origin of Species*, and indeed the intellectual lineage can be traced much further back than that. For instance, Augustin de Candolle in the early 1820s remarked that "all plants and animals are at war with each other." Tendrils of intellectual descent can even be traced back to Plato, who posited what is referred to as the "principle of plenitude," which holds that all stations of nature are filled. The complete filling of these stations in turn indicates that there would be some competitive force maintaining and defining the position or role of each species. As Adrienne Mayor described in her superlative book, *Fossil Legends of the First Americans*, several Native American tribes of North America also invoked what we would today describe as biotic explanations for extinction, whereby the remains of certain Pleistocene fossil mammals they discovered in their natural environment were posited to have been driven to extinction by humans. (We interpret this as a biotic explanation for extinction because competition with and predation by humans was the culprit).

This is not to say that competition doesn't exist. Indeed, it is a force manifest in nature (and human endeavors). It just doesn't seem to be what most of the time causes what we refer to as macroevolutionary change: speciation and extinction. We both love nature shows, including the eponymous *Nature*, as well as others presented by various august sources including *National Geographic*. One of the fascinating competitive dramas that has been well documented (albeit anthropomorphically dramatized) is the set of interactions between hyenas and lions on the grasslands of Africa. These interactions served as key features of the plot of the classic animated flick from 1994 *The Lion King*, which features a cavalcade of stars doing sublime turns as voice actors, including James Earl Jones, Matthew Broderick, Whoopi Goldberg, and Cheech Marin (of Cheech & Chong fame). The (mostly) noble lions, excepting the evil villain Skar sublimely voiced by Jeremy Irons, are pitted against the drooling and (mostly) ignoble hyenas. Treachery, battle scenes, and other biotic interactions are pursued to great effect in the movie. The interactions and the behaviors on the one hand nicely mirror some of those displayed by the animals in those nature shows yet are very human at the same time. One gets the sense from the nature shows that lions and hyenas genuinely dislike each other, though we realize that we too may be anthropomorphizing as much as the film does. Still, to see alpha male lions slay alpha female hyenas in the real world not just for meat but for the simple pleasure of it and the associated disruption of the hyena pack that follows, and to observe packs of hyenas set on individual or multiple lions and kill or attempt to kill them, again with no interest in killing for meat, is fascinating and disturbing.

When Bruce queried Elisabeth about this very same competitive drama in the wild and whether some might view it as counterevidence to her turnover pulse, she patiently said "not at all" and then went on to explain why. She remarked that lions and hyenas have co-occurred in Africa for the last three million years; during that time there have been no sustained directional trends in any aspect of their anatomy. In essence, lions and hyenas perfectly epitomize stasis and punctuated equilibria, and morphologically they have not changed one whit over the course of all those interactions played out down those millions of years. There was much interaction between members of these two species, but when it came to their macroevolution, there was little to none.

As a sidebar, Bruce regrettably never got the chance to ask Elisabeth what she thought of *The Lion King*, but even more regrettably he did once recommend that she watch the TV show *Beavis and Butthead*. With some postdoctoral advisors such tomfoolery could have amounted to professional suicide, and regrettably Elisabeth did follow this viewing recommendation for the first and likely last time that she put any stock in Bruce's TV/movie reviews. Upon viewing it she remarked, twice, that it was "awfully silly," in her proper British accent and with a greater emphasis on the "awful" part. Fortunately, Elisabeth was a stellar individual who withheld judgment when it came to an advisee's predilection to sometimes watch lowbrow TV shows.

CHARLIE BROWN, LUCY, AND SISYPHUS: THE MYTH

Turnover pulses, while not frequent on the timescale of human existence, are commonplace geologically, with many dozens occurring over the Phanerozoic eon, the last 540 million years, which represents the fountainhead of animal and plant evolution. More rarely, but still at least five times in the history of animal life over the last five hundred or so million years, there have been even more cataclysmic mass extinction events, involving the rapid elimination of species at a global scale. As we have mentioned throughout this book, these mass extinctions were a phenomenon that quite frankly Steve Gould was infatuated with, for he recognized their implications for understanding evolutionary theory and the history of life. Intriguingly, some mass extinction events had causes that were extraterrestrial in nature, such as the giant meteorite crashing into the earth in the Cretaceous, and one may have been triggered by a star exploding outside our solar system. Why does this matter for our understanding of evolution? The existentialist in us declares that "nothing matters," yet we persist in plugging along. Maybe life is like that too? It continually gets smacked down, yet at least thus far has inexorably "come back for more." It is like the Sisyphean drama that Charles Schulz so masterfully scripted in the ballet between Lucy, Charlie Brown, and that most highly renowned of oblate spheroids, the pigskin of American football. Charlie was if not inordinately fond at least kind of fond of kicking footballs through field goal uprights, unable to resist the temptation to do so. This was used to advantage by fellow *Peanuts* cartoon denizen Lucy, who was somewhat immodest and occasionally unkind. Periodically viewers would be confronted by a scene with Lucy, shown balancing a football between the ground and her forefinger, and Chuck, steeling himself and then sallying forth into a run to dramatically win one for the team by kicking the ball through the uprights. Yet we all knew that at the last millisecond, at the very end of Charlie's run up to the ball and swing of the leg, that Lucy would snatch the ball away, inevitably sending Chuck flying, to land back first some distance away. It hurt physically, but it hurt even more emotionally.

Darwin closed his magnum opus with a masterful quote: "whilst this planet has gone cycling on according to the fixed law of gravity, from so simple a beginning endless forms most beautiful and most wonderful have been, and are being, evolved."[2] But he could have equally closed it with the more lugubrious phrase: "forms most beautiful and most wonderful have been and are being driven to extinction."

We suspect that *On the Origin of Species* is much more uplifting than *On the Death of Species*, though in our modern world the latter may be timelier. However, Darwin's vision of the origin of species very much involved one species as it originated, competitively wedging out another species that had previously existed, given that he held that there were a limited number of stations of nature and thus resource space available, revisiting Plato's principle of plenitude. Consequently, and emphatically, in Darwin's worldview, evolution and extinction were intertwined and could perhaps be seen as two sides of the same coin. This was carried even further by the aspect of Darwin's worldview, and the neo-Darwinian worldview, that emphasized the notion of what Niles and Steve termed phyletic gradualism, whereby species would slowly and gradually evolve into the station of one species and out of the station of another species, what is called phyletic speciation conjoined to phyletic extinction.

We aren't saying that this aspect of Darwin or the neo-Darwinian view is valid. In fact, it is very much debated whether there is a cap on the number of species that can persist in any one setting or on the globe in total. Further, it is not generally accepted that the evolution of one species requires another species to be competitively driven to extinction: rarely does one species competitively drive another species to extinction during any portion of its history. (See our earlier discussion of charismatic, and less charismatic, carnivores on the African savannah and *The Lion King*). Last, for our present purposes, we only need mention that the phyletic gradualist perspective is supported by little evidence: our own favorite quote on this is Darwin's quietly desperate note on the lack of evidence supporting his early embrace of gradualism (1837–39) in one entry in one of his *Transmutation Notebooks*: (paraphrasing) . . . surely Mr. Lonsdale has a few examples. . . . Mr. Lonsdale, manifestly, never showed up with the goods. One intriguing possible instance of gradualism was discovered by Dana Geary, paleontology professor at the University of Wisconsin, and one of many female graduate students advised by Stephen Jay Gould; she, and colleagues, focused on fossil clams from ancient lakes in central and eastern Europe and did find that while most of the species examined showed stasis, a few displayed gradual change over time.

Instead, today while it is well recognized that speciation and extinction events appear to be clustered in time, as Elisabeth Vrba recognized with turnover pulses, the causal mechanism driving this clustering is not competition but rather environmental change, which in turn influences the geographic range of species and their likelihood of going extinct or producing new species.

What causes this environmental change? To a greater or less extent, the answer often lies in astronomy. Particularly in the case of the turnover pulse, Milankovitch cycles are governed by the dynamics between our sun and the changes in the amount of solar radiation the surface of Earth receives due to changes in the orbit, rotation, and axis of our planet. These cycles occur over the course of tens to hundreds of thousands of years. We note that there is excellent evidence of these preserved throughout the geologic record. For instance, consider the Devonian of upstate New York State.

THE DEVONIAN OF NEW YORK,
WHICH DESERVES TO BE A MOVIE

Like most New Yorkers, we define "upstate" as that portion of New York that extends from the northernmost tip of the Bronx to the Canadian border. Like most geologists, we consider the Devonian a first-class geologic period. The Devonian is named after the Devon region of England, where it is well represented and was first documented by British geologists. However, arguably, if the constituent rocks of New York state had been examined first, then at least by the criteria of temporal completeness, range of environments, and geographic spread it could well have been called the New Yorkian—and indeed was almost called the Erian, in honor of the easternmost of the Great Lakes. But we will not fall into the jingoistic trap that we castigate in chapter 11. The Devonian of Devon and England is assuredly first-class too. Suffice it to say, though, that there are lots of Devonian rocks, and fossils, in New York. The Devonian in general is known for preserving several key episodes in the history of life: the first fossil forests, coincidentally also from upstate New York; the proliferation of sharks and other fishy things in the oceans; and the first vertebrates to sally forth on land for consistent parts of their life span.

When it comes to the narrower confines of the state, paleontologists also record many spectacular Devonian trilobites from the region. These include *Eldredgeops rana*, which we mentioned in chapter 2 and served as a touchstone for Niles (when it was *Phacops*) as he discovered the key evidence for punctuated equilibria, as well as several species of *Bellacartwrightia*, which Bruce named after his wife, Paulyn Cartwright.

The region also contains places of transcendent beauty, extolled by many, including the exponents of the Hudson River School of Art. Parts of this real estate, which we and especially Niles prospected in search of trilobites, were blanketed very long ago in the Devonian by new and reworked sediments, the detritus from a mountain range whose uplift was triggered by the collision of fragments of ancient continental plates with North America. It was in this area that geologists Peter Goodwin and E. J. Anderson, formerly of Temple University, identified "punctuated aggradational cycles," which they suggested were caused by Milankovitch-driven oscillations in sea level.[3] (These cycles are preserved in oozy lime muds in the segment of the Devonian deposited before that continental collision.) Although the virtues of the punctuated aggradational cycle or PAC hypothesis may not have been widely espoused, we thought it was a pretty cool idea back in the day. Although it has not yet been examined, it would be interesting to see if these Devonian climate cycles produced corresponding pulses of speciation and extinction, like Elisabeth Vrba's turnover pulses preserved in the latest geological intervals of the African fossil record. If they did not, it may have been because the Devonian denizens of the Catskills were marine invertebrates such as brachiopods and trilobites that were less likely to become geographically isolated and speciate than Elisabeth's tropical mammals. Most marine creatures, especially animals like clams and brachiopods, speciate more slowly than tropical mammals, and this is just a manifestation of the volatility phenomenon we described in chapter 6. In fact, there *are* major turnovers seen in Devonian invertebrates, coming at less frequent intervals than PACs, and roughly every five million years at stage boundaries. This indicates that there are scale differences in turnovers.

GONE IN 60 SECONDS

Other types of more calamitous "astronomically driven" environmental changes happen more rarely. There is of course the giant meteorite of end of Cretaceous fame. Sometimes a star across the galaxy explodes in a gamma ray burst (GRB), releasing a pulse of energy lasting thirty seconds or so, and thus gone in circa half the time of the car thievery feats celebrated in Nicolas Cage's movie *Gone in 60 Seconds*, the merits of which we might consider challenging save for the fact that Nicolas loves trilobites almost as much as we do. The amount of energy released by a GRB is truly prodigious, equivalent to the energy produced by the light of all the stars in the universe, and the most powerful explosions in the universe.

Based on research Bruce conducted in collaboration with physicist Adrian Melott of the University of Kansas and colleagues, a GRB 440 million years ago and 6,000 light-years away may well have played a key role in causing the mass extinction at the end of the Ordovician period.[4] This mass extinction spelled the end for perhaps 70 percent of all marine species, including many representatives of the group we know and love, the trilobites. For instance, never again would the charismatic species *Homotelus bromidensis* from the Ordovician of Oklahoma (figure 8.1) scurry across the seafloor.

Whether a GRB precipitated the end Ordovician mass extinction (partly by triggering a profound but brief episode of global cooling) is not yet settled. However, understanding by astrophysicists of past and present rates of star formation and explosion suggests that a GRB should have occurred at least once in the last billion years in our galactic vicinity (about 6,000 light-years away, give or take). We can also be fairly certain that there was never a proximate GRB, let's say 500 light-years away or less, over this same interval because it would have incinerated the face of the planet and wiped out life entirely, the ultimate mass extinction that thankfully never came to fruition.

Citizens love paleontology, and when you combine fossils with exploding stars it represents a subject matter that earnest folks everywhere find hard to resist. Indeed, when our hypothesis on a GRB and the late Ordovician mass extinction

8.1 *Homotelus bromidensis* specimens from the Yale University Peabody Museum of Natural History. Photo by Bruce S. Lieberman, used with permission and courtesy of Tim White, Derek Briggs, and Susan Butts.

first was published Bruce received a phone call from a sincere employee of some western state's Department of Fish and Wildlife enquiring about the threat GRBs represented to various birds of prey. Thankfully, his concerns were allayed in short order, and the stewards of California condors can sleep peacefully at night knowing that a cosmic catastrophe in a spiral arm of our galaxy is far down on the list of perils confronting endangered representatives of the class Aves.

JAWS

Another more recent and different type of stellar explosion also seems to have left a signature in the geological record as well as a stamp on the history of life: a supernova. About 2.6 million years and 150 light-years away a supernova exploded; intriguingly, and perhaps not coincidentally, this is when Vrba's most pronounced turnover pulse occurred. This prodigious explosion may be associated with a decent-sized extinction event in some big animals, but since there weren't that many big animals that doesn't translate into a mass extinction. Supernovas are less violent than GRBs, releasing only one-hundredth the energy, but that's still 10^{44} watts of power. Because they are less violent, and their explosive energy is released outward in a dispersed blast radius rather than concentrated, directed beams, faraway supernovas do not comprise the collective cosmic threat to us that GRBs do. Geologically recent and astronomically nearby supernovas can, however, leave behind discernible geochemical signatures such as the radioisotope Iron-60, which GRBs do not, making their occurrence in deep time hard to detect. It turns out that an Iron-60 spike is present in 2.6 million-year-old marine sediments. Adrian Melott and colleagues helped draw the connection between that spike in the sediments and a closely corresponding time horizon when several large taxa disappeared from the fossil record, most notably the charismatic megafaunal *Carcharocles megalodon*.[5] The types of radiation released by a supernova would be most detrimental to organisms with large body size, and the Megalodon, a kind of ancestral shark, was as big as a bus. It certainly was also voracious, awe-inspiring, terrifying, and the most fearsome marine predator of its epoch or perhaps any other. It may have been one of the inspirations for Steven Spielberg's brilliant film *Jaws*, and it certainly was the inspiration for action hero Jason Statham's 2018 substantially less brilliant but still moderately diverting vehicle *The Meg*, where even he was hard pressed to survive his encounters with the beast.

Rosa Compagnucci of the Universidad de Buenos Aires and colleagues also argued that a supernova at this time may have not only triggered the extinction of Megalodon via the radiation it released, but such a starburst would have led to the extensive climate cooling observed at 2.6 mya; that in turn appears to have triggered extinction in other taxa. We note that other phenomena, including the closure of the Isthmus of Panama due to the collision between North and South America, have previously been suggested as the major cause of the switch to the icehouse conditions that characterized the Pleistocene epoch and commenced at this time, but wouldn't it be fascinating if a distant exploding star also played a role in this change? We think so. Melott and Brian Thomas of Washburn University further described some of the other effects a nearby (astronomically speaking) supernova might have on Earth, including an uptick in lightning strikes and concomitant fires.

Of course, there is a problem with all of this: the elephant in the room, or giant shark in the swimming pool, is that sometimes we have no clue as to the precise reason(s) why some long-extinct species failed to survive longer than it actually did. Connecting Megalodon's demise to a supernova certainly veers into the realm of the speculative, although such speculation could inspire future tests of the connection between the birth and death of species and stars. (It may be even harder to connect Megalodon's expiration to a supernova, given that some argue the beast disappeared before the event in question.) When it comes to mass extinctions, the confluence of many extinctions does tend to strengthen the case for a connection to some overarching factor or factors. And the extinction event 2.6 million years ago was certainly not a mass extinction, according to paleontological consensus.

All science is about progressing from finding correlations to developing causations, and the Megalodon/supernova example is certainly still in the "finding correlations" stage, meaning this idea has not progressed far beyond infancy just yet. It may well never do so. The sun may just be another star, but it's our star: as Joni Mitchell once sang, "We are stardust." May it never explode.

CHAPTER 9

Is Eternal Sex Necessary?

Or, What Are These Coywolves Doing in My Backyard?

Abeautiful four-acre public woodlot adjoins Niles's backyard in northern New Jersey. It's great for migrating birds in May, and dog walkers and teenage beer-drinking boys near the end of the school year. Even the occasional bear.

And, often, white-tailed deer. There are many of them, arriving as solitary two-year-olds hopping the fence so easily, munching down on Niles's beloved garden flowers, or as mating pairs, or as mothers and their little ones. Once Niles saw a trio of stags with fresh new antlers, all gussied up for the mating season. In short, his yard sported pretty much all the known social-ecological permutations and combinations of white-tailed deer.

Lately, the only known predator other than human hunters and car drivers to pose a threat to this suburban explosion of venison-on-the-hoof has appeared as well: a wild doglike species definitely not known in the greater New York City environs when Niles and Bruce were kids seventy and fifty years ago, respectively, or, for that matter, when Niles first moved to his present location in the mid-1980s. People were calling them "coyotes," but, with the exception of one of Niles's first sightings, these animals seemed to be larger than the western coyotes he had seen over the years in the east. They tend to hang out in small packs of up to six or seven individuals. Probably family groups, Niles thought, trying to sort out the voices on a night of nearby howling.

After the local press started covering them, photos of actual encounters and not just "file photos" appeared. One is indelible in Niles's mind's eye: an apparently dimorphic pair standing over the carcass of an adult deer, killed out on the ice of a lake only ten miles or so from Niles's house in New Jersey.

9.1 Two coywolves born in the wilds of Massachusetts. Photo courtesy of Jonathan Way of Eastern Coyote/Coywolf Research from his book *Coywolf*, available at https://www .easterncoyoteresearch.com/downloads/Coywolf.pdf, used with permission.

Coywolves are beautiful (figure 9.1). They are hauntingly reddish, mixed in with and surrounding the brown, black, and silver hairs of their flanks. They look very much like wolves. Not coyotes, though coyotes are a sort of wolf, as are domestic dogs, all members of the genus *Canis*. The pair illustrated in Niles's local paper, standing beside their fallen prey, looked like actual, real live wolves, thrilling to see in the suburbs of New York City.

Niles met one of them one May morning in 2017. He was sitting in the wood lot on the trunk of a storm-fallen tree—his habitual perch when he wants to sit, think, dream—and see whatever is there crawl, walk, or fly by. Niles noticed movement to his right, where the ground dips down to a little stream. The head showed first, and Niles first thought it was one of the neighbors with his chocolate lab. Neck and shoulders appeared next. No collar—and no neighbor holding a leash. And with a finer, pointier face, maybe not quite as large as the dog he expected. He, Niles thinks, was a uniformly, gorgeous reddish brown all over. Then the full body appeared as the cautious approach continued. He was looking for something. He didn't seem to know Niles was there.

He started walking slowly along one of the trails, taking him in Niles's direction. He paused occasionally to look around, but soon drew opposite Niles, by then perhaps fifteen feet away. And suddenly he took a left and started right toward Niles. Niles said a soft "hello" when he was still eight or ten feet away. The coywolf stopped and looked Niles straight in the eye, and after a second or two turned left again and started walking unhurriedly back to the stream.

By that time Niles had done some desultory searching of what these newly arrived wolflike dogs could be. He thought of the red wolves, considered by many biologists the rarest, most endangered wolflike canid in North America. Red wolves, Niles found out, have been protected since 1987 in a small area near the coast in North Carolina. They are known to be hybrids between western coyotes and eastern timber wolves, with some domestic dog genes mixed in. They are smaller than eastern timber wolves but larger than western coyotes. They live in small packs and take down deer as well as a variety of smaller prey. And they are distinctly reddish.

Niles's first, nonexpert but distinctly curious guess was that maybe the conserved red wolves were doing so well in the southern Appalachians that they were beginning to expand their range northward. Further north even than the New York area, as Niles first heard them, not in his backyard in New Jersey but on forested ridges, howling away in the dark of night in the Adirondacks. People there started calling them "coy-dogs" back in the 1990s. Niles had heard that a small pack killed a white-tailed deer at the boat launch at a little pond where his family has a summer cabin.

Rummaging through the digital amalgam of newspaper clippings and some recent papers in the scientific literature, Niles was glad to see that some evolutionary biologists had taken notice of this rather sudden appearance of an otherwise unknown wolflike predator in our human-crowded exurban/suburban—and even just plain urban—midst. And Niles quickly learned that most biologists who had begun to compare the genes of these "northeastern coyotes" (their newly minted albeit informal name) with eastern timber wolves, western coyotes, and yes, with the red wolves in their diminished numbers living down south, saw them as something distinct from the rest. Something new.

It was mind-blowing. Here we had the sudden presence of a new canid, a form not known until just after World War I, when some were seen in southeastern Canada in 1919. There are some coyote pelts in the Blue Mountain Lake Museum in the Adirondacks of northern New York—but they, like their larger eastern timber wolf brethren, had been "extirpated," shot and trapped into oblivion, the way people often treat wildlife that poses some sort of perceived threat directly to themselves ("big bad wolves") or to their livestock or wild species that humans have an interest in. After all, people still eat deer.

That was a century and more ago. But now, suddenly, "they" seemed to be back. Early results of genetic testing showed that these latest arrivals were also hybrids between western coyotes and eastern timber wolves, plus a variable presence of domestic dog mixed in for a perfect trifecta. Niles began thinking that the smallish one he had locked eyes with and that came to his backyard off and on for a couple of years might have a fair amount of suburban dog in him, the more so because he was also accompanied by an apparent sibling that had abnormally short legs. That pair were probably responsible for the killing of at least two fawns Niles knows of. When Niles had that encounter with the larger of them on that fine May morning, Niles thought he was out scouting around for a lone female wild turkey that had been coming into Niles's yard every day, eating the cracked corn he had laid out. It was odd that she was always alone, odd that she never was with any other wild turkeys. At that time there seemed to be no other small flocks of turkeys in the immediate vicinity. Niles had seen her the day before, but he never saw her again.

Most arresting of all was what Niles learned from this small handful of initial scientific papers he found on the internet: whatever these newly appeared coyote-like animals were, they were *not* actual red wolves. Red wolves, it turns out, were also the product of some hybridization event in the past. But they have different proportions of wolf and coyote (and dog) genes in them: a stamp of a past genetic, hence evolutionary, hybridization event. And, looking at these newcomers, you can see they are too small to be exactly "the same" as the red wolves. But the data, and the various biologists who were collecting and analyzing, pretty much agreed that these modern, newly arrived "coywolves" (as some began to call them in the late 2010s) were *newly evolved*.

Can you imagine that? An entirely new kind of large mammalian predator evolving essentially before our eyes? In a matter of less than a hundred years, from an initial few after World War I, populations have spread up and down the eastern seaboard of North America—from eastern Canada, throughout New England, down south at least as far as Virginia, westward through West Virginia and Pennsylvania into Ohio.

Some biologists think that these "coywolves" of the northeast are already a distinct species. Dogs (the family Canidae, that is) are notorious for their hybridizing pretty freely with other dogs, whether domestic dogs (think of forms running the gamut from Chihuahuas to Irish wolf hounds) or what seem like separate natural species such as western coyotes and various forms of "true" North American wolves. Wolves will kill and eat coyotes, but they will also mate with them. And when they mate, their offspring are often, probably even usually, viable, functional, and reproductively fertile.

Yet it is not all genetic and anatomical chaos out there. Wolves, to the extent there are any significant numbers left, tend to breed with other wolves. Packs have their own ranges. Likewise with western coyotes. There is more than a semblance of adaptive genetic, phenotypic (anatomical and behavioral) order to the groups of canids out there. Hybridization essentially offers a pathway to faster-than-usual evolutionary events. But it does not obliterate, as a rule, the tendency of distinct and typically stable "adaptive types" to form in discrete regions and comprise small bands of similar-looking, similar-behaving populations, whether or not they are truly reproductively isolated from similar groups elsewhere—the hallmark of true species in evolutionary biology. (Here we must add, though, that a few centuries of such stability do not necessarily stack up to the stasis measured in the hundreds of thousands and even millions of years we see in the fossil record.)

In short, it sure looks like what we have had at least for the past century or so is the rapid formation of a new ecological type of wild canids that look and behave very much like an earlier one that still clings to existence elsewhere. Seemingly, though, the younger one is not derived directly from the older one. Rather, they are both products of different hybridization events from the same antecedents: western coyotes and eastern timber wolves.

Evolutionary reinvention of the ecological wheel. Or so it would seem. Evolution before our very eyes. Rapid evolution. Amazing.

As soon as it became clear from reading the few papers Niles could find that these coywolves in his backyard were (1) not here a hundred years earlier and (2) not *anywhere* a hundred years earlier, and as soon as Niles realized they were something new, an image popped up in his mind: a picture of a polyhedron, a multifaceted three dimensional ball-shaped object with many sharp-edged symmetrical sides, as if someone carefully sliced off pieces from the surface.

And then it came to Niles: Stephen Jay Gould had written about that image as used by an earlier biologist. Niles remembered, thankfully, that it was near the end of one of the twin papers Steve had published in 1980 in one issue of *Paleobiology*, at that point a still fairly new journal dedicated to the spirit of encouraging more active theoretical thinking in paleontology. The two papers had caused quite a stir. Both were a clarion call to action: "The Promise of Paleobiology as a Nomothetic, Evolutionary Discipline," and "Is a New and General Theory of Evolution Emerging?" Taken together, juxtaposed as they were, they made quite a splash. They also annoyed a hefty percentage of paleontologists and—especially the latter one—evolutionary biologists in general.

Niles quickly found the polyhedral passage he had remembered from the second article. It was in the summation/exhortation to move forward in the more general piece, the one describing the "modern synthesis" of evolutionary theory in the throes of being replaced by a more complete, hierarchically structured theory—a theory that sees evolution as embracing more than just allelic frequencies changing or remaining the same under pressures of mutation, drift, and (especially) selection in the evolutionary process. Higher-level entities, most especially species, were to be seen as actual players in the evolutionary game. For this was the paper in which Steve, with his already famous audacity, assured all who cared that the modern synthesis "is effectively dead." The charge still rankles in many quarters all these years later, though the rebellion has yet to be fully realized, even though much progress has been made.

Steve was especially concerned with bringing back organisms into the evolutionary discussion. To that end, he quoted the biologist St. George Mivart, the man who had had such an up and down relationship with the idea of evolution, as well as with Darwin himself. Steve, writing of what he calls Galton's polyhedron, quotes Mivart quoting Darwin's half-cousin Francis Galton:

This conception of such internal and latent capabilities is somewhat like that of Mr. Galton . . . according to which the organic world consists of entities, each of which is, as it were, a spheroid with many facets on its surface, upon one of which it reposes in stable equilibrium. When by the accumulated action of incident forces this equilibrium is disturbed, the spheroid is supposed to turn over until it settles on an adjacent facet once more in stable equilibrium.

The internal tendency of an organism to certain considerable and definite changes would correspond to the facets on the surface of the spheroid.

Steve adds, "under strict selectionism, the organism is a sphere. It exerts little constraint upon the character of its potential change; it can roll along all paths." He goes on to note that, seen as a faceted polyhedron, we enter the world of

constraints exerted by the developmental integration of organisms themselves. Change cannot occur in all directions, or with any increment; the organism is not a metaphorical sphere. When the polyhedron tumbles, selection may usually be the propelling force. But if adjacent facets are few in number and wide in spacing, then we cannot identify selection as the only, or even the primary control upon evolution. For selection is channeled by the form of the polyhedron it pushes, and these constraints may exert a more powerful influence upon evolutionary directions than the external push itself.[1]

Steve loved metaphors. He was always somewhat bemused, back in the day, when Niles told him that one of his two favorite papers among Steve's earliest productions was "Eternal Metaphors of Paleontology" (1977), on the directionality, cause, and tempo of evolutionary change. The other was "Is Uniformitarianism Necessary" (1965), a play on Thurber's "Is Sex Necessary?" In the second, Steve argued that the geological maxim of "uniformitarianism" associated first with James Hutton, but most especially with the later Charles Lyell, embraces two separated albeit related thoughts: the first being that geological processes we observe going on today are the same as those that were operating through all time, producing all the changes in the earth we can decipher in the rocks; and the second being that such processes are invariably and uniformly slow and gradual. The first, Steve said, is true enough, but reduces quickly to simple induction; the second is simply false. Niles thought both papers were inherently wise and thoughtful, and their maxims and observations have stuck with both him and Bruce over the years.

(It is worth noting that Steve, to our knowledge, never published his reflections on Thurber's original question. We note that sex is the only known physiological process not essential to the survival of a living eukaryotic organism, an observation that has nontrivial valence in figuring out what life is all about, as developed to some extent in other essays in this volume. But try telling that to a teenager. Yet, as far as *eternal* sex is concerned—everybody will agree that the answer is yes, for without it, an animal species would simply cease to exist. And that is of course true for humans as well.)

Steve went on to say a bit more about the polyhedron metaphor, if only in passing: "Most of the other changes in evolutionary viewpoint I have advocated throughout this paper fall out of Galton's metaphor: punctuational change at all levels (the flip from facet to facet, since homeostatic systems change by abrupt shifting to new equilibria)."[2] "Punctuational change at all levels" would seem to include the "punctuated equilibria" model itself—though, curiously, Steve does not actually spell this out. But as we pause over the thought that Steve was taking a metaphor rooted in organismal phenomenology and hand waving it to apply "at all levels," we point out that he was in very good company: several of the key architects of the neo-Darwinian synthesis had done the same. For instance, Sewall Wright with his "adaptive landscape" (first published in 1931 and 1932) initially portrayed "harmonious gene combinations" on "adaptive peaks," but later expanded his thesis to embrace entire (conspecific) populations and even species occupying the peaks. Thaddeus Dobzhansky (1953) took that imagery and said the evolutionary history of sexually reproducing organisms entails Wright's adaptive peaks and valleys, but also entire "ranges" of closely related larger-scaled taxa; George Gaylord Simpson (1944) conceived his "quantum evolution" to explain the often abrupt appearance of higher taxa, such as orders of mammals like bats and whales, but said the process goes on at all levels, including species, where it would be very close to punctuated equilibria (only lacking the important component of stasis). Ernst Mayr (1963) followed suit, but in a more restrained way.[3]

Steve and Niles last worked together (apart from a few "where are we now?" short papers) in 1977, when, in part, they dealt with the apparent reality of species and their putative role as evolutionary agents as a necessary implication of punctuated equilibria.[4] Niles had suggested in 1979 that "taxic" evolution could be usefully distinguished from microevolution involving selection and drift within local populations, and that speciation and extinction rates might be determined to some degree by the width of ecological niches.[5]

But it was the following year, 1980, when a small but nonetheless significant burst of macroevolutionary explorations appeared independently. In addition to Steve's twin set of papers, along with Joel Cracraft, Niles started exploring other macroevolutionary patterns (for example, adaptive radiations) in the context of phylogenetic systematics—the approach to reconstructing biotic history in terms of purity ("monophyly") of descent developed by German entomologist Willi

Hennig (1966) that revolutionized the practice of biological systematics. Joel and Niles recognized the existence of different phenomenological levels (e.g., populations vs. species), each with their own internal evolutionary processes in some sense and to some degree "decoupled" from one another, as suggested by Steven Stanley in 1975, and indeed in the original punctuated equilibria paper of 1972. And they emphasized the nature of taxa as packages of transmissible (that is, genetic) information that of course has great implications for understanding the nature of the evolutionary process itself.

Perhaps most significantly, Elisabeth Vrba made her appearance as well on the published world stage of emerging macroevolutionary theory, taking the discussion to a new level with her antelope data and analytical analyses. Among other things, Elisabeth introduced her "effect hypothesis," where the adaptive attributes of organisms could account for much of the patterns (e.g., evolutionary change) seen among species within clades preserved in the fossil record.[6] She concluded that "the sole motor ultimately driving evolutionary change seems to be the environment," a prelude to her work a few years later on turnover pulses, a critical central facet of the evolution of life. That year, 1980, was a big one for pushing the macroevolutionary ball/polygon down the road. (More on these concepts can be found in other chapters in this book.) Elisabeth wrote one or two papers with Steve. Niles wrote one with her. But he never worked directly with Steve again.

Back to Galton's polyhedron, the surprising explosive evolutionary appearance of coywolves after World War I in eastern North America, and how the two might be related. For whether these eastern coywolves are truly a new reproductively isolated species (and, after all, as prone to hybridization as canids in general are, it is unlikely), these novel eastern coywolves are of a familiar "adaptive type." Their behavior and appearance call to mind a comparison with the earlier hybridization event that produced red wolves. In April 2022, Niles had the privilege of listening in to an online lecture at the University of California Santa Cruz, given by Princeton biologist Bridgett vonHoldt, an expert on red wolves and a student of the ecology, behavior, and evolution of North American canids more generally. Though focusing on red wolves and the survival of their genes in canids in Texas, vonHoldt did mention in passing the "red wolves of the northeast." And so they seem to be, a sort of reinvention of the same sort of canid as the dwindling red wolves.

It is as if, Niles has been thinking, a second hybridization event landed on the "same" or a closely adjacent facet of Galton's polyhedron: a reinvention of an adaptive type of wolf, smaller than existing eastern timber wolves, but large enough to prey on eastern white-tailed deer. Steve spoke of the internalized organismal "constraints" of these facets, but it is just as apt to think of them as opportunist "adaptive types," the configuration of phenotypes best suited to succeed in the ambient (and highly degraded) environments of current North America—from exurbia to the inner cities themselves. These similar phenotypes (again, in behavior as much as anatomy) could be produced by similar, if not identical, identities and proportions of the components of the underlying genome.

The rapidity of the evolution of this red wolf-like "morphotype" is mind-boggling. The idea of punctuated equilibria is based in part on the intransigent stability, the "stasis" displayed by most species throughout their known durations. When change comes, it comes *fast*. But *how* fast is fast? Niles speculated back in the 1970s that maybe we were talking about five thousand to fifty thousand years, a "guesstimate" pronounced by some population geneticists as actually being quite slow, since natural selection can and does act much faster than that. A refreshing reminder when he first heard it. But here we have a new large-bodied adaptive type—possibly a full-fledged new species—appearing and developing in a hundred years. That looks like a great example of the rapid change called for in punctuated equilibria.

Steve held out this metaphor, this Galton's polyhedron, as a possible component of a new, extended, hierarchical evolutionary theory. Niles wondered if he had missed something: What had Steve ended up doing, specifically with reference to this metaphor? Had he explicitly connected the polyhedron metaphor to punctuated equilibria in some paper somewhere? Niles felt sure that he must have done so. It seemed to him such a perfect fit. It still does.

Steve fell desperately ill in 1982. He said he was very tired when he, Elisabeth Vrba, and Niles got together with colleagues at a conference in Dijon, France, that May. His next few years were devoted largely to beating back the cancer that had taken hold, though he was justifiably proud that he never missed a monthly deadline for his famous *Natural History* magazine essay series "This View of Life" during the ordeal. When he started to recover to the point where he could resume his scientific work, he dove into Cambrian paleobiology, emerging with his book *Wonderful Life* (1989), with its interpretations of diversity and disparity and the roles of chance and "necessity" (homage to Jacques Monod) in the history of life. Roll the tape of life again from some point—say, the Middle Cambrian—and you will see a different outcome from what we have around us today. True enough, though Niles has always been drawn more to the repeated evolution of similar adaptive solutions to life's varied, yet oft-repeated challenges as ecosystems are continually reassembled after disturbances ranging in magnitude from localized fires and volcanic eruptions up through the occasionally global devastations of bolide impacts, major eruptive events, and the like. Similar "adaptive types" seem to be fashioned by the evolutionary process, over and over again.

Steve took great pains to disassociate himself from what he called "naïve adaptationism." He never told Niles if he considered him to be a "naïve adaptationist," and Niles hopes that his adaptationism is not, in fact, naïve. But he does think that punctuated equilibria, at heart, is a theory of adaptation, melded with a taxic perspective: the origin of discrete species. It says that adaptive evolution is not, as a rule, slow, steady, and gradual. There are periods in and around the origin of new species that typically bring adaptive change to the descendant species—sometimes a lot, sometimes hardly any at all, most times, in true Goldilocks fashion, just the right amount to suit the new species to somewhat different environmental conditions than its ancestor occupied—or still occupies. Naïve or not, Niles is in that very real sense an "adaptationist."

Steve doesn't mention Galton's polyhedron in *Wonderful Life*. Nor does he elsewhere use it in any other of his papers, as far as we can remember or otherwise discover. So it was natural to look for it, as a last resort, in Steve's final statement on the nature of things—his monumentally long book *The Structure of Evolutionary Theory* (SET, for short). If memory serves Niles, it was sometime in the early 1980s when Steve first told him that he had begun writing his equivalent of Darwin's "Big Species Book." But the book was written desultorily (as its repetitious nature would seem to imply)—and lingered in boxes in his administrative assistant's office, gathering dust until the news of the second wave of Steve's cancer set off a frantic sorting and writing episode that she described to Niles at Steve's memorial service at New York University. The book was published in 2002, the year Steve died.

At first glance, there was joy in seeing a moderately long list of page references to Galton's polyhedron in the index to SET. But disappointment soon followed: rather than link the metaphor to speciation and specifically to punctuated equilibria (let alone any further extended evolutionary theory), Steve confines it to discussions of saltation, orthogenesis, and other mooted internal controls of change in early discussions of evolution. An exception occurs near the very end, where Steve presents a *different* quote from the same St. George Mivart, which does indeed make the expected connection. Mivart says, as Steve writes:

> Arguments may yet be advanced in favor of the view that new species have from time to time manifested themselves with suddenness, and by modifications appearing at once (as great in degree as are those which separate *Hipparion* from *Equus*) the species remaining stable in the intervals of such modifications: by stable being meant that their variations only extend for a certain degree in various directions, like oscillations in a stable equilibrium. This is the conception of Mr Galton, who compares the development of species with a many faceted spheroid tumbling over from one facet, or stable equilibrium, to another. The existence of internal conditions of animals corresponding with such facets is denied by pure Darwinians.[7]

So Mivart saw punctuated equilibria and linked it to Galton's polyhedron. But Steve chose to dismiss Mivart's comments here as purely "saltationist." Yet it is nonetheless a good description of the basic punk eek pattern. And for some reason, Steve chose not to point that out.

Niles tends not to think of Steve as the full-blown anti-adaptationist he chose to present himself as. In a footnote on page 397, Steve excoriates himself for his "strict" adaptationism as a young professional. Yet he was right to develop his own takes on the roles of chance, the constraints (and opportunities) channeled in development, and the apparent randomness of catastrophic events (random at least in regard to adaptations, if not to the laws of physics).

In the end, we two, who knew him so well and respected him so highly, believe that Steve would not disagree with the more measured take on adaptation through natural selection as the core evolutionary paradigm developed first by Darwin and coming down in variously modified forms, not only in the evolutionary biology of sexually reproducing organisms but also in somatic evolution (immune system, cancer—maybe even in the brain of individual organisms) and sociocultural evolution in all its various manifestations, from stone tools to complex institutions. All involve transmissible information whose fate is decided via a form of selection in an economic context. That's what makes them all "evolutionary."

So Steve's resurrection of the polyhedron mooted by Galton and developed by Mivart has real potential in thinking about evolution as we know it today: evolution involving stasis and change of transmissible information. Its products may well often be reliably stable adaptive phenotypic states. In its explicitly eco-evolutionary context, and in tribute, we propose the metaphor be termed "Gould's polyhedron."

CHAPTER 10

Darwin in the Galápagos

Running the *Beagle* Tape Backward

T he Galápagos Islands have been on the UNESCO list of World Heritage Sites since 1978, partly in recognition of the fact that Charles Darwin arrived there in August 1835 and saw and collected many of the elements of the local fauna. He left with a conviction, albeit hidden in his private notes, that life had evolved. Darwin had been thinking so since before he got aboard the *Beagle* in late December 1831, or so we think. He'd certainly been flirting with the idea—again, privately, in his voluminous notes and a couple of essays now tucked away in the *Earthquake Portfolio* in the rare book collections of Cambridge University—since the *Beagle*'s first stop in the Cape Verde Islands.

Darwin had been in South America for nearly four years, and the Galápagos were his final stop. He was eager to get there, and in a letter written in Peru he predicted to his cousin "Fox" what he hoped he would find there. Darwin had seen, collected, and recorded his thoughts, and he mostly saw the sorts of things he had thought he might see. As the *Beagle* set forth on its homeward voyage across the Pacific, Darwin wrote in his *Ornithological Notes*, "If there is the slightest foundation for these remarks the zoology of Archipelagoes—will be well worth examining, for such facts [would] undermine the stability of Species."[1] Darwin inserted the word "would" as an afterthought, lest Robert FitzRoy, the *Beagle*'s captain, read his notes before they reached the safe shores of England. Historians squabble to this day about when Darwin "became an evolutionist." But these words from Darwin are the earliest evidence, to (nearly) everyone's satisfaction, that the Galápagos experience sealed the deal in Darwin's mind.

The Galápagos didn't give him the idea. He had read his grandfather Erasmus's decidedly evolutionary book *Zoonomia* as a kid. In medical school in Edinburgh he had read the works of the early transmutationists Jean-Baptiste Lamarck and

Giambattista Brocchi. He had also studied with the evolutionist Robert Grant, as well as with Robert Jameson, who introduced Darwin to the works of Georges Cuvier—himself mute on the subject of transmutation, but an influential pale-ontologist who saw the waves of extinction that had periodically wiped out ancient species that were then replaced by younger ones.

Later, at Cambridge, having failed out of medical school, Darwin studied botany with John Henslow and geology with Adam Sedgwick, both ordained clergymen and creationists in heart and mind. Yet these two nonetheless gave Darwin the practical keys to studying variation within species in the field and the lab (Henslow) and how to decipher the historical sequences of layered rocks (Sedgwick). Darwin certainly knew about transmutation and what you might expect to find in the field if evolution in fact was a natural process underlying the history and diversification of life on earth. Darwin pulled out of English waters on the *Beagle* in late December 1831 with a prepared mind on matters transmutational.

The Galápagos are justly famous in the annals of evolutionary biology—and thus in the annals of civilization, since evolution is an idea that continues to enrich and disturb the hearts and minds of humanity. Darwin, again, did not invent the concept. But he surely established it on a sound intellectual, scientific footing—all the while fearful of the explosion of skepticism, some of it pure hatred, that he well knew his work would trigger.

But what about other places—such as Down House in Downe, Kent, south of London—where Darwin lived with his wife Emma and a passel of kids from 1842 to his death in 1882? He wrote the *Origin of Species* there, and all his subsequent books as well. He took his ruminative walks on the Sandwalk. He did a lot of experimental botany in his purpose-built greenhouse out back behind the main house. Yet, though Down House is recognized by English Heritage, it has not made the grade to full UNESCO World Heritage status, despite spirited attempts to have it so declared in 2009, the two hundredth anniversary of Darwin's birth.

And what of other crucial sites that were demonstrably important to the for-mation of Darwin's evolutionary ideas? How about the two places along the shores of Bahía Blanca in Argentina, where Darwin first arrived in the fall of 1832, three full years before he got to the Galápagos? Darwin reveled in finding fossils, comparing them with still-living species in the vicinity, drawing his conclusions about extinction, stasis, and ancestry and descent of the vertebrate and inverte-brate species as he compared the fossils he found with the still-living species, a

10.1 Circa twelve-thousand-year-old human footprints alongside the huge, deep tracks of a *Megatherium*, a bipedal giant ground sloth, with the former possibly hunting and about to kill the latter. The site is near the shoreline of Bahia Blanca, Argentina. Photo courtesy of Teresa Manera, used with permission.

crucial early set of observations underlying Darwin's early embrace of the very idea of transmutation.

Those beachside exposures of sedimentary rocks at Bahia Blanca are now augmented by something spectacular that Darwin did not see: massive numbers of footprints of extinct giant sloths and other species, interspersed with human footprints in a layer dated at about twelve thousand years before the present (figure 10.1).

There is another, still younger layer, about seven thousand years old, that had lots more human footprints but no trace of the larger, now extinct fauna except the footprints of a large bird that may well have been left by the greater rheas still living in the vicinity, poignant if mute testimony to the drama of human arrival in the New World and the havoc we caused as the existing megafauna were fairly quickly driven to extinction. Darwin did express concern that humans were already posing a threat to the wildlife of South America. Why not recognize this already locally conserved beach site along the shores of Bahia Blanca, where the *Beagle* had anchored on two separate occasions during its epic voyage, as a balancing act to the Galápagos, as a World Heritage Site that celebrates Darwin's evolutionary ideas as well as the modern predicament of the massive, human-caused "Sixth Extinction" engulfing the world's biodiversity at this very moment?

Why not, indeed. Niles returned, after an eye-opening earlier trip in 2008, to attend a conference in Argentina devoted to this very issue: exploring ways and means of successfully nominating the Bahia Blanca sites as worthy of UNESCO World Heritage recognition, so far with no more success than the Down House nomination.

During a lunch break at the conference, vertebrate paleontologist and historian Paul Brinkman, sitting next to Niles, asked if he thought Darwin would have come up with the very same set of ideas about evolution had the *Beagle* gone the other way around the world, meaning that the Galápagos would have been his first port-of-call in South America. That was Stephen Jay Gould's rolling-the-tape-over-again question with a major twist: If Darwin saw the same sites he saw on his three-year sojourn in southern South America, *but in the reverse order*, would he have come to those same ideas on evolution that he did in real time?

What a terrific thought experiment! What if Darwin had indeed seen the Galápagos first, then mainland Peru and Chile, with a round-trip side venture up and over the Andes to Mendoza, Argentina? He would then have gone down to Valdivia, Chile, then the Chiloe Archipelago, and thence through the Magellan Straits to spend two years plying up and down the Atlantic coast of Argentina, making depth measurements to improve navigation charts to support the military and commercial interests of Great Britain in the era when many South American countries were fighting for their independence from Spain. Darwin frequently made treks both large and small overland—to see what was there, of course, but also to get off the ship where he was perpetually seasick. (His remit had originally been to look for exploitable mineral resources—which he failed miserably to achieve). Then, finally, he would have gone north to Rio de Janeiro and Salvador da Bahia de Todos os Santos in Brazil before heading out to the Cape Verdes and back to England, where he would have finished his *Red Notebook*. Would he then have begun his four famous *Transmutation Notebooks* and said much the same things in them all? More simply, would he even have written that sentence in the *Ornithological Notes* where he said that his observations in the Galápagos shed doubt on the "stability of species?"

Think about it: Paul's gambit was to wonder aloud if evolution were to start all over again from a certain point, would we see the same resultant outcomes? More bluntly, would we be here, sentient bipedal chimpanzees who learned to walk and talk, and maybe even think? Who embrace music and the arts while occupied as well with killing each other off, the better that our own "special interest groups" (family, friends, neighborhoods, ethnic identities/regional loyalties/entire nation states) can hog up whatever blandishments the deeply injured natural world still proffers to our greedy hands?

Ray Bradbury already answered his version of this question in the negative in his famous 1952 story "A Sound of Thunder," about a Mesozoic butterfly being killed by a time traveler—a hunter on a "Time Safari" seeking to kill a *Tyrannosaurus rex*, the ultimate trophy. Bradbury suggested that the butterfly's death, which would not have occurred at the time and in the manner that it did if humans had not time-traveled back to that time and place, would have had less drastic consequences: humans would be here, but they would be spelling words differently. The results of the recent election were changed, and the autocratic "strongman" won. Given what we face at home and elsewhere right now, these are somber words indeed.

Our present question on Darwin is a bit more specific. It addresses the history of thought of a single human who was adept at, even insistent about, keeping his thoughts to himself. He interviewed people and read assiduously. He seems to have spoken or written to no one on the *Beagle* about transmutation. Darwin's thoughts were formed, some discarded, some kept, as his journey progressed—and as his journey in life in general continued to progress after he got home in 1836. He periodically added to his thoughts on evolution by steps, some larger than others.

It will spoil nothing about the rest of this chapter if we say right now that Darwin's early thinking on transmutation pretty much precluded the entire concept of adaptation, even though the whole world, including a large number of modern professional evolutionary biologists, thinks that evolution—meaning Darwin's evolution, but also including what they see as evolutionary thought today—is really all about adaptation. The lesson of the Galápagos is still nearly universally seen as a story of adaptive evolution in the beaks of what have long since been called "Darwin's finches"—more properly termed simply "Galápagos finches." We agree (as we have made clear elsewhere in these essays) that adaptation is indeed at the core of the evolutionary process. But Darwin's actual work on the *Beagle*—in its entirety, but certainly on the Galápagos—was primarily about the replacement of closely related species in space as well as in time. It is a signal fact that no one back in the day (except Lamarck, the vastly underrated and demeaned natural history polymath in Paris) saw transmutation as even involving adaptation.

Transmutation initially was primarily about the replacement in time of extinct species by younger descendant species, a topic for those naturalists who were conversant and comfortable with the fossil record as well as the living world surrounding them. It was Darwin, as we are about to see, who added species replacement in space as well as in time. Up through the rocks—or over the surface of the Earth. Darwin's title, in 1859, was *On the Origin of Species*—and we have already seen that Ernst Mayr and many other biologists said that the title was a weird misnomer, since Darwin doesn't actually speak of the processes of speciation as it had come to be known by the mid-twentieth century. He basically has the modern concept of species as reproductive entities, and of geographic speciation, in his *Transmutation Notebooks* of 1837-1839, especially *Notebook B*, Niles's nominee as the intellectually most important single item of Darwin's voluminous writings. None of these items were published until the late twentieth century.

When Darwin did inject adaptation into the mix, in that *Notebook B*, he had come to reformulate the central problem of evolution to be the understanding of why organisms tend to match up so well with the external world to fulfill their basic economic and reproductive needs in life. The problem was the diversity of phenotypes, rather than species per se. He saw natural selection as necessarily incremental, based on small heritable variations—and he saw that it was good. Species, as a concept, had to go: they weren't at all "stable" but kept changing with the vicissitudes of time and space. He clung to the title "Origin of Species" as an inherited mantra announcing the topic of transmutation, hanging on as if to an atavistic organ no longer serving the use it was originally intended for.

But back to the tape: Steve meant the actual details ("tape") of the biological, evolutionary history of life. The metaphor holds too to the historical development of a chain of thought about something like evolution, whether in one person's head or in a babbling throng of individuals discussing politely with or angrily denouncing one another as more is learned about what life, past and present, actually has been and continues to be, at least for the nonce, and as the details of life's history come into sharper focus, as bodies of modern and ancient life forms are supplemented by the tools of molecular biology and biogeochemistry.

This suggests that thoughts themselves evolve. Yes, it's history. But old ideas are discarded when they are no longer useful—because they no longer seem to be "true," to paint a plausible picture of the Way Things Are, at least as we now think they are. This is ideational evolution. New ideas come to take their place; there are variations in those ideas, and they are selected—by different individuals, and by entire groups who tend to fight over them. In fact, everything that humans manage to do other than basic physiological, behavioral, and of course biologically physical traits that evolve according to standard rules of biological evolution (however we understand them to be at the current moment) are done via conscious thought. Cultural evolution is just as real as biological evolution, and many times more complex.

"The evolution of Darwin's evolution" is no empty phrase. And when we roll the "Darwin on the *Beagle*" tape backward and forward, we are running a tape of his ideas on the processes underlying life's history, and not his ideas (also evolving) on what the details of life's history have actually been. Bradbury's tape was something else yet again: it was the trip back through time, accelerating from a slow start, taken by a fictional set of guides and hunters as they went from the present back to the Upper Cretaceous. (Bradbury in 1952 dated that period to sixty million years before the present; we know now that it had to have been minimally sixty-five million years.) Darwin's trip through South America was through a bit of time (four years), but it was mostly through space. We trace his ideas evolving in a context where time is important (sequence of events), but where space is governing the order in which he saw things, just as it was Darwin himself who added space to time as his observations and theories evolved.

Niles's answer to the historian's provocative question, after a moment of surprise, was "no," as the historian knew it would be. Why? There were things Darwin needed to see first to comprehend the significance of what he would see in the Galápagos in August 1835. For instance, in January 1832 the *Beagle* was quarantined at the Canary Islands due to a cholera outbreak back home in England. Rather than wait out the twelve days, Captain FitzRoy traveled to the Cape Verde Islands. Darwin beachcombed on Quail Island and found some marine fossils indistinguishable from the cast-up shells of living animals littering the strand line. He wrote in his diary: "To what a remote age does this in all probability call us back & yet we find the shells themselves & their habits the same as exist in the present sea." Darwin was discovering what we have termed throughout this book "stasis."

In February 1832, when he arrived at Salvador da Bahia de Todos os Santos in northeastern Brazil, he wrote "The Mind is a Chaos of Delight"—expressing joy and amazement at the immense variety of different species, as well as the riot of colors, in the Atlantic tropical rainforest. Darwin got another chance to experience that "Chaos of Delight" sensation later in the fall, when the ship pulled into Bahia Blanca, Argentina. He found more fossil shells that once again appeared to be the same species living in the nearby shallow water. He also found bones of fossil mammals, including giant ground sloths and glyptodonts (huge versions of armadillos). These had previously been recorded by Georges Cuvier and other European naturalists. Though extinct, their close relatives—tree sloths and armadillos—were still very much alive.

What thrilled Darwin most was his discovery of fossil bones of a small rodent at Monte Hermoso. He thought these belonged to a smaller species ancestral to the living mara, or Patagonian cavy. Maras are the third largest species of rodents alive today. Only the capybara, living in the lusher more tropical parts of South America, and the American beaver are larger. At first glance maras look more like giant bunnies than any rodent most of us have ever seen. Niles was sorry to learn a few years ago that they have made it into the pet trade and are now facing extinction.

Darwin started his trip focused on the replacement of fossil species by living descendant species. The "mara" he found as fossils was basically the same as those alive today, only smaller. Richard Owen described these fossil bones in the 1840s, in a monograph done under Darwin's overall editorial direction. He found them to be the bones of an ancient chinchilla, not a mara, although there are actual Mara skulls in those same fossil beds. They look pretty much like the living species but are sufficiently different to warrant recognition as a separate species—so Darwin was right, if not for precisely the right reason.

His next step was to search laterally over space rather than through time. Darwin started comparing wildlife species as he made his way overland. Crucial to that was a stop in Maldonado, Uruguay, in 1833. The weather was bad, the crew was grouchy, and there were few if any rocks to prospect for fossils. Darwin had some crew members collect birds, and he became a pretty good field ornithologist. He started seeing patterns of replacement of closely related species as he traveled around the far reaches of Argentina. At times in his notes he would say that species *Y* here "takes the place" of species *X* there.

The most striking example of this pattern of geographic replacement involved the two distinct species of rheas—the ostriches of South America, smaller than their African relatives, but still rather large flightless birds. The greater rhea (ñandú) was common over the pampas of the northern half or so of Argentina, and Darwin and the *Beagle*'s painter, Conrad Martens, would shoot rheas for the ship's dinner on occasion. They surely would have seen them at Bahia Blanca the first time they were there. Niles did in 2008, though by then they were much less common and hard to find. Darwin had heard, as the ship kept going back and forth along the southern Atlantic coast, that there was a second, slightly smaller and browner rhea species—the choique—that "takes the place" of the ñandú in the Patagonian scrublands.

It took a while, but he finally found one. It was on his dinner plate on Christmas Day, 1833. He had felled a guanaco and Martens a small rhea, which Darwin assumed was an immature ñandú. He says he sat up like a shot when he realized what he had been gnawing on. He raced off to the galley too late to save most of the carcass—but rescuing a wing, the legs, the head and neck, and some large feathers. And he went on to see the choique in the field. They are still fairly common on Peninsula Valdes (figure 10.2), where Niles was lucky enough to run into three groups of males herding their chicks around, just as male dinosaurs may have done seventy million years before.

Darwin learned that the two species meet along the Rio Negro, the boundary between the pampas to the north and Patagonian scrubland to the south. They meet but they do not mate. In Darwin's terminology, they are "inosculant," literally meaning they don't kiss, a suggestive term but also a formal way of saying the species do not in any sense grade into one another. They remain quite distinct even when they meet. In modern terms, they don't interbreed and are "reproductively isolated."

THE IMPORTANCE OF ARCHIPELAGOS

Darwin upped the ante in his two trips to what the British call the Falklands and the Argentinians angrily correct to the Malvinas. These are islands off the Atlantic coast of southern Argentina that happen to contain gorgeous Devonian

10.2 A view of one of the Miocene cliff faces making up Peninsula Valdes, Argentina. Photo by Niles Eldredge.

trilobites. Niles was invited to join an expedition to collect these that never took place because of the war that broke out between Argentina and Great Britain in the early 1980s. And he was hissed and booed in a lecture he was giving in Ushuaia, billed as the southernmost city in the world, that was otherwise very well received, because he stupidly referred to the "Falklands." Whatever you call them, these were islands visited by Darwin in 1833 and again in 1834 that had a formative effect on him.

There are two main islands in the Malvinas/Falklands—East and West. A sealing captain, a Mr. Lowe, had told Darwin of an animal he called the "Falkland Fox." He said that they were on both islands and that they were distinctly, consistently different from one another. Bearing that in mind, Darwin collected two from one island and three from the other. He did so, as we saw in chapter 6, by walking up to one and crushing its skull with his geological hammer. He wrote in his notes that such passive, unwary behavior would soon lead to the extinction of these canids. He was of course right. They are no more, victims of the spread of modern humanity.

Darwin agreed with the captain. The two islands' fox populations were indeed consistently different, though he saw the two as nonetheless constituting a single species. They seemed related to species of foxes or wolflike canids known on the South American mainland.

Darwin next encountered an archipelago when the *Beagle* arrived at Chiloe along the coast of southern Chile in early 1835. But it was in the Galápagos, reached later in August of that year, the culmination of nearly four years of probing and thinking all over southern South America, where the light bulbs really started going off—though not quite in the way the textbooks tell it today.

Darwin wrote his cousin William Fox from Chile, telling them how excited he was to be going to the Galápagos. He said that he hoped there would be Tertiary aged rocks so he could find fossils to go along with the modern plant and animal life. In that, he was to be disappointed: there were no substantial sedimentary rocks on those rather barren, desolate, pretty much waterless volcanic islands (figure 10.3). But his mind was prepared to find what he did find: species related to mainland South American species yet different, and with luck species that differed from island to island, like those Falkland/Malvinas foxes.

Now is the time where you might expect our narrative to turn to finches. Darwin detested them. Had it not been for his helpers in the *Beagle* crew, whom he had trained to be assiduous collectors, none would have been collected. Darwin personally did not collect any of them. The finches presented no neat packages of distinct, easily sorted-out groups. They were instead a confusing mélange of variation to Darwin's eyes, both on individual islands and across all the islands. Only long after he had arrived home did he learn from ornithologist John Gould (no relation to Steve) that there were indeed tolerably discrete clusters of finches in thirteen species, still pretty much the view of modern ornithologists. Only then could Darwin see their value—using the finches in his 1845 second edition of the *Voyage of the Beagle* to hint that all those finches seemed to suggest a derivation from some common ancestral form.

The mockingbirds (mentioned in chapter 5) and tortoises gave him what he was looking for. Mockingbirds are yet another group native to the Americas. Most of the living species are in the tropics and down through South America. The thenca is the mainland Chilean mockingbird, distinct from those species that Darwin saw on the other side of the Andes in the pampas and in Patagonia. Darwin saw that there were four distinct species living in the Galápagos, though one was so similar to the mainland thenca that he noted only a difference in song.

10.3 Santa Cruz Island, Galápagos Islands. Photo by Bruce Lieberman.

When John Gould analyzed the birds that Darwin and crew collected on the voyage, he agreed with Darwin's assessment—and the intervening years have shown little change in the taxonomy of the Galápagos mockingbirds. One species occupies most of the islands. But in the southeastern, geologically oldest islands there are three more, each on separate islands. One is nearly extinct. Each, as Darwin said of the Falkland Islands foxes, is "consistently" different from those living on nearby islands in the archipelago. We repeat here what Darwin said in his *Ornithological Notes*, written on his way home and reminiscing about the Falklands as he writes about his mockingbird and tortoise observations: "If there is the slightest foundation for these remarks the zoology of Archipelagoes—Will be well worth examining; for such facts [would] undermine the stability of Species."

Darwin came to see the geographic patterns of replacement as halos: the larger halo for, say, mockingbirds were the Americas as a whole; then the Chilean coast plus the Galápagos mockingbirds, with those on the coast related to but distinct from those on the islands; and, on a still smaller scale, evidence of differentiation on various islands within the archipelago.

And, so, back to that historian's question, again the answer is that Darwin would not have had the same experiences, and thus the same development of thought, if he had reached the Galápagos first. There were no fossils to find and interpret as possible ancestors to living species. Without his prior experience seeing geographic patterns and then extending them to offshore islands, he might not have been equipped to see that pattern immediately there, and certainly not ready to have that a-ha moment to declare that species are derived from one another. Darwin had spent nearly three years training himself to be a decent field ornithologist, without which he might have been oblivious to that point.

In summary, before he reached the Galápagos Darwin showed up knowing what his elders had said about transmutation. The main theme was this: if evolution occurred, then modern species must have descended from and then replaced now-extinct species in that climb through time, one best seen, naturally enough, with fossils, especially invertebrates, whose remains are typically so abundant and easy to find and to collect. In the very first shore adventure, in the Cape Verdes, Darwin found, instead of change, obdurate stability, stasis. Modern species were not created yesterday.

Six months or so later, Darwin next discovered what he thought was the fossilized remains of an extinct species ancestral to the huge modern rodent, the mara. A-ha! It was the first major observation corroborating what the fossil record should look like if life had evolved, at least the way Giambattista Brocchi saw things, if not exactly the way Jean-Baptiste Lamarck did. Replacement of species through time, first exemplified by the four charts that Darwin's mentor Robert Jameson had published, utterly without comment, in his fifth edition of Cuvier's *Theory of the Earth* in 1825.

In 1833 Darwin and his band of helpers, partly for want of anything better to do, started in earnest a shipboard collection of dead birds when the *Beagle* was tied up in Maldonado, Uruguay. It became the basis of Darwin's extension of the original "replacement in time" gambit with the novel "replacement of closely related species in space." At first, Darwin seemed comfortable with a combined spatiotemporal replacement of species in his evolutionary thinking. Only later, after he had arrived home, did apparent either/or style conflicts between the two—time and space—arise in his thinking. (But that's another story. See Niles's *Eternal Ephemera* for more on all of this, from soup to nuts, including Darwin's post-*Beagle* thinking on up to the publication of *On the Origin of Species* in 1859.)

On the *Beagle*, the next step came when Darwin learned of the "consistent differences" between the two closely related fox species on the East and West Islands of the Falklands/Malvinas archipelago. Islands seems to generate forms different from their relatives on the mainland, and even different among the islands within the chain.

Darwin had all that under his belt before he ever got to the Galápagos. He was so steeply versed in these fundamental perspectives of species replacement in space and time that he was able to predict what he in fact ended up seeing when he wrote his cousin William Fox.

Let's assume that the governor of the Galápagos would have told Darwin about the different shape of the tortoise shells on the different islands if Darwin had showed up there before seeing the rest of southern South America. Would that have been enough to seal the deal? It's doubtful that he would have noticed the differences between the mockingbirds on the different islands.

Evolutionarily speaking, Darwin might very well have been forced to be content with fulfilling his first recorded wish early in the trip: to write a book on the geology of South America. That he came back with theories of mountain building and coral atoll formation and indeed wrote up the geology of South America and compiled important collections of natural history specimens and observations might have been enough. But a lot of the initial enthusiasm for the subject might have dimmed by the time he got to see the whole picture. The Galápagos was a magnificent confirming, predicted experience, best appreciated by Darwin's already prepared mind.

CHAPTER 11

Of Cultural Nationalism, Hamlet, and the Cloaca Universalis

Why Citation Is the Best Policy

No one likes to get ripped off. Whether it's a song, a joke, a recipe or, as in the story we are about to recount, original scientific data and interpretation, no one likes it, and all the more so if it potentially costs you money. A colleague once pointed out that in the academic arena, pay seldom matches that of former schoolmates who went on to generally more lucrative pursuits in law, say, or medicine or business. All we scientists and our fellow "hewers of wood and drawers of water" in academia, to quote that same colleague, have is the satisfaction and reputation among our colleagues that we have done good, intellectually valuable—and, to a degree at least, *original*— work. And though we regularly publish our work, in doing so we want and expect our names to be "forever" associated with our work. ("Forever," when it comes to having a museum hall or a hospital wing named after you may turn out to be seventy-five years or so, if we consider examples from modern New York City representative.)

In other words, we expect to be properly cited. If money is your goal, becoming a paleontologist is not your best bet. Still, we all have to make a living, and at times even us fossil workers find the evocative track from mega-spoof band Spinal Tap "Gimme Some Money" cycling through our brains.

The real rewards lie in the thrill of the hunt and the discovery and collection of fossils. It is way cool to find a piece of once-living history, especially when it sheds further light on the evolution of life and the factors that shaped it. Fossils are their own reward to those who cherish them.

Unlike genetic information, which is transmitted primarily vertically from organisms to their descendants (except in microbes, which can at times transmit genetic information horizontally between evolutionarily distinct microbes that just happen to be living together), cultural information is transmitted both vertically and horizontally. News, both fake and real, travels ever faster, and theft of ideas is rampant in all walks of life. Patent law was invented to slow if not completely stymie stealing ideas in the marketplace, though the barriers presented by patents themselves often lead to still further inventive creativity, which is a good thing. Cultural evolution is more complex than biological evolution, and it is often faster, too, at least when we compare the visible adaptive changes to the things that people make to the way people themselves are biologically configured.

But theft of idea is indeed thievery, no doubt about it. Intellectual property rights are currently a major concern on the global economic stage. It was ever thus, as our story of the "Hamlet Affair," which unfolds in 1841, makes abundantly clear.

Our protagonist is James Hall (1811–98), born in Massachusetts two years after Charles Darwin was born in England. He was the first paleontologist of the State of New York. Yes! States had state paleontologists back then, and in some instances they still do, for reasons of exploring the natural resources of the territory, above and below ground, especially if not solely for potential economic benefits.

James Hall was one of the first and finest of what has long since developed into a world-class (world-leading, actually) American tradition in the geosciences, including what we now call paleobiology, geobiology, or biogeology. Hall worked on the great *Geology of New York* (1836–41), a geological survey of four separate geographic regions ("districts"). The work was authorized by the state legislature under a still larger and more ambitious undertaking entitled *Natural History of New York*. The geologists were charged with mapping the rock outcrops in New York and determining the sequence of sedimentary layered rocks in the state. Hall was in charge of the Fourth District in western New York, embodying a spectacular sequence of Lower and Middle Paleozoic rocks (Cambrian–Devonian) (see chapter 8 for more about the Devonian), many of them jammed with fossils.

Hall was appointed state paleontologist in 1841, and he soon associated professionally with people such as Ebenezer Emmons, who was said to have had such a profound disagreement with James Hall about the age of rocks in the Taconic Mountains of New York State that he was compelled to leave the state entirely. Hall eventually went on to publish thirteen volumes of the *Paleontology of the State of New York*. Nearest and dearest to our hearts is the famous volume on Devonian trilobites and other arthropods published with his assistant, John M. Clarke, in 1888. This volume, and the others, can be purchased used for a hefty fee. They were sumptuously bound in the acme of nineteenth-century bookbinding

11.1 John M. Clarke in his Victorian-era office at the New York State Museum. Courtesy of New York State Museum and Brad Utter, used with permission.

splendor, featuring images of trilobites and other ancient marine creatures on their covers, embossed in golden foil.

It was the Paleozoic rocks and fossils that Hall studied, illustrated, and described, that aroused the greatest interest, both at home and abroad. For our introduction to this tale, we can do no better than turn to John M. Clarke's (1857–1925) thoughts on the subject. Clarke (figure 11.1) was Hall's successor as New York state paleontologist—and his biographer in 1924 as well. There is a famous picture of Clarke later in life holding a bell, which the National Academy of Sciences tells us was rung by a "galvanic current sent through a mile of copper wire" that Clarke managed to procure for the New York State Museum's collections.

In its day, Clarke's biography of Hall was highly acclaimed. The august geologist Thomas Chamberlain remarked that he doubted whether there was "any other text that carries the reader so close home to the inner history of our own and allied sciences in this country during the early and middle stages of the last century. No student of geological and paleontological progress should miss the opportunity to read this thesaurus of information on a most vital stage of early

American science."[1] Granted, "the inner history of our own and allied sciences in this country during the early and middle stages of the last century" sounds suspiciously like "profound decisions were made by the great American presidents William Henry Harrison and Zachary Taylor."

In any event, Niles recognized the significance of this document as part of a broader debate on culture and nations. When Clarke got to what was later called the "Hamlet Affair," he began with these forceful words on the early days of our nation and its nascent developing efforts in science:

> The volcanic outburst of reports from the Geological Surveys throughout the States from 1835 to 1845, years when the sovereign competency of the States was unchallenged and national consciousness was not fully awake, aroused wide interest in Western Europe. We must stop to remind ourselves that geology was still a very young science just emerging from a nebula of hypotheses and contentious guess-work into an orderly and rapidly increasing array of concrete cause and effect. Its novelty, the tremendous sweep of its propositions and the romance of its buried treasures gave all its adventures wide popular appeal. The English and French geologists were making rational progress in laying the foundations of historical geology and with Italy and Switzerland were finding out the principles of dynamic geology.
>
> The new-found developments from the western world were therefore of exciting interest, and as soon as the reports of the New York men were spread abroad there began an invasion of the country by European geologists who would compare the old world with the new and help to set the whole terraqueous globe in order.[2]

In other words, in the early 1840s the United States was at peace, and scientific developments were burgeoning in the country. Some scientific discovery was supported at the individual state level, and on the geological front knowledge was already pouring in in the form of huge monographs. New York led the way, with reports on all four of its designated geographic divisions completed by 1845. In contrast, the U.S. Geological Survey was not formed until 1879.

Yet the justifiable pride in the high-quality work of the early American geologists engaged in sentient exploration was not limited to just geology, or indeed confined to any given state. Americans were well engaged by the 1840s in founding ever more institutions of higher learning, soon to be followed by all the other

11.2 The interior of the Galerie de Paléontologie et d'Anatomie compareé (Gallery of Paleontology and Comparative Anatomy, Paris, France). Photo by Bruce S. Lieberman.

institutional symbols of European-inspired higher culture: parks and gardens, zoos, museums of art and science, halls for concerts and operas. The Academy of Natural Sciences, for example, was founded in 1812 in Philadelphia. Natural history museums are our personal favorites of all such cultural institutions, and we have enjoyed visiting natural history museums all over the world (figure 11.2).

Science can be a culturally nationalistic phenomenon. Sometimes this can spur positive developments, like the circumglobal proliferation of natural history museums. Other times it can have negative connotations—for instance, the colonialistic and/or imperialistic practices of some of the scientists at these museums, who pillage the natural history treasures of another country for personal, professional, or financial gain. There can also be haughty or dismissive attitudes or actions displayed by scientists from one country toward or against scientists from another country. These positive and negative aspects partly center on the collection of and access to data. But cultural nationalism in science can even extend to the currency of ideas and hypotheses. It is to one such instance that we now turn our attention: an episode in the history of science where British scientists tried to tell the geologic story of the United States and American scientists metaphorically fought back to try to exert some degree of ownership. Although nowhere near as portentous as the dramatic "shot heard round the world" fired in Lexington, Massachusetts, in 1775 that set the American Revolution ablaze, it did represent a formal scientific stepping-out party, with American geology and paleontology emerging from the penumbra cast by European science in general and British science in particular. Afterward, the United States would become an engine of geological and paleontological insight on the world stage.

Natural history museums are educational in nature. Ever since the American Museum of Natural History in New York City was founded in 1869, it became a mecca for kids, some of whom were destined to become medical doctors or scientists. But all natural history museums are dedicated as well to amassing large behind-the-scenes collections of specimens and to supporting a highly trained scientific staff to curate and to perform original research on them.

Pride individual and national was on the line when the foreign invasion of eager European scientists began their travels to the newly minted United States. They were awaited on our shores with a mixture of eagerness and anxiety: eagerness to show off our marvelous rock sequences and fossils, as well as our own homegrown geological and paleontological human prowess, anxiety inasmuch as the Europeans were the unquestioned heavyweights while our scientists, our best and brightest, were as newly minted as a group as their nation. Among the Europeans, the British loomed the largest, our former colonizers, from whom we had not all that long ago wrested our freedom. Britain, with its much longer and distinguished history of doing science, boasted geologists who were known worldwide.

Foremost among these famous British geologists is the antagonist of our piece: Charles Lyell, who made his first of four visits to our shores in 1841. Lyell was, and remains, best known for his three-volume *Principles of Geology* (1830–33). He was a prolific writer, regularly continuing to revise his *Principles* and his later *Elements of Geology* (1838) as well as producing numerous papers and other books in his long career. By the time he arrived in America he was forty-three years old and long since world-famous. He was an inspiration and later a close friend to Charles Darwin, who was twelve years Lyell's junior. Lyell was one of the small circle of colleagues and relatives to whom Darwin revealed his ideas on evolution in 1844. Lyell was very sharp and knew the literature—but, it seems, he was too quick to bow to the prevailing political winds. In our opinion, he well knew that the evidence for evolution was compelling even before Darwin showed him it was so, but Lyell did not accept evolution publicly until after Darwin published his *Origin of Species* to instant acclaim in 1859, owing, we think, to faintness of heart.

The old boys were not as scrupulous about citing their sources in the first half of the nineteenth century as we are encouraged to be today. They usually did mention their predecessors, but who was there, really, to cite before them when the game of science was so utterly new? Lyell is best known for his championship of uniformitarianism, the principle that the "present is the key to the past," meaning that we can explain geological history with references to the processes we can observe going on above, on, and beneath the earth's surface today (a.k.a. "actualism"). As Stephen Jay Gould pointed out in one of his earliest papers in the 1960s, actualism is commingled to varying degrees with the notion that things usually happen slowly and gradually and that the earth is in something of a steady state, with little or no directionality to it as the eons of time roll slowly by. In the sense of slow steady gradual change, uniformitarianism stands in contrast to catastrophism—in paleontological circles, at least, associated with the great French geologist/paleontologist Georges Cuvier and his "revolutions on the surface of the globe." Cuvier saw ancient life as a succession of more or less stable biotas, periodically eradicated by catastrophically destructive events, to be replaced by successor biotas by processes he declined to explore—at least on paper. The justly renowned vertebrate paleontologist George Gaylord Simpson once somewhat sarcastically characterized his American Museum of Natural History colleague Norman D. Newell (see chapter 7) as a "neocatastrophist" when Newell renewed the Cuverian theme of "crises in the history of life." Notably, Steve was also often characterized in this manner, for he saw rigid adherence to uniformitarian principles as not only inaccurate but also standing in opposition to broader acceptance of punctuated equilibria. Steve believed, moreover, that key tenets of the neo-Darwinian synthesis that evolutionary change was always slow and gradual, which were partly derived ultimately from uniformitarian principles, intellectually stunted evolutionary biology. This perspective led evolutionary biologists to devalue certain types of data, especially from the fossil record, which were not so much in line with the principles of slow and steady change. Steve saw a direct line leading from uniformitarianism to stultifying attitudes toward paleontology. Intriguingly it was Simpson, the key paleontological figure of the neo-Darwinian synthesis, who very much emphasized how evolutionary rates were not constant and gradual but rather vary substantially through time (notwithstanding his derogatory labeling of Newell).

The notion of uniformitarianism that comes down to us today is something of a mish-mash. It is still always attributed to Lyell, who indeed developed and popularized it. But he got it from the great Scottish thinker and geologist James Hutton from the previous generation of British scientists. According to Robert Silliman, Lyell's failure to acknowledge Hutton's earlier work adequately caused widespread displeasure among other geologists.

Lyell was trained in the law and was an intellectual dilettante, with no formal standing in academic circles. Darwin eagerly awaited the arrival of the second volume of Lyell's *Principles of Geology* when his travels aboard HMS *Beagle* brought him to Montevideo, Uruguay. Some have observed that volume 2 reads like a legal brief. It is essentially a summation and refutation of Jean-Baptiste Lamarck's theory of evolution. It is undeniably brilliant, for it is an accurate characterization of the brilliant Lamarck's views. There seems to be no outright chicanery in Lyell's writings, but his observations manifestly failed to convince the young Darwin. Darwin was already thinking of transmutation based on the fossils he had collected at Bahia Blanca, Argentina, in September–November 1832, just before going to Montevideo. And, of course, as the *Beagle* journey wore on, Darwin became more and more convinced of the simple, ineluctable fact of evolution. For our money, we think Lyell's big contribution in volume 2 of the *Principles* was, rather surprisingly, a convincing grasp of fields we today call ecology and biogeography.

Lyell was also the first to subdivide what was then called the Tertiary period into four epochs: Eocene, Miocene, Lower Pliocene, and Upper Pliocene. He did so using the percentage of clam and snail species found in sedimentary units that are still alive today. Lyell does acknowledge the data of the French paleontologist Gérard Deshayes, but he did not mention that Jean Baptiste Lamarck in 1801 had concluded that only some 3 percent of the mollusks found in the sedimentary rocks around Paris are still alive today—rocks that fall into Lyell's Eocene epoch. Nor did he mention that Giambattista Brocchi in 1814 concluded that nearly 50 percent of the molluscan species preserved in the sediments of the Apennine Mountain foothills of Italy are still with us today. These would lie in Lyell's Lower Pliocene. In other words, the idea of calculating the percentage of fossil species that are still alive today and using that as a guide to determining the age of a fossil deposit came from work by important scientists conducted a generation before Lyell published his *Principles*. Furthermore, the works by Lamarck and Brocchi in which they described these patterns were definitely known to Lyell.

Charges of plagiarism were to dog Lyell throughout his career, at home in Britain as well as abroad. Perhaps the most pronounced charges occurred in the 1860s and concerned his *Antiquity of Man*, Lyell's belated embrace of evolution. But already by the 1840s Lyell's reputation as a thief of the data and ideas of others was already well established. According to Silliman, a British geologist told Hall that Lyell "has the same reputation here as he does abroad, and his cognomen is *Cloaca Universalis*." The universal sewer, indeed.

It is rather disconcerting that someone today often celebrated as the "father of geology" was in his own time recognized as a blatant plagiarist (and called a universal sewer, for goodness sake, not even just any old sewer but a universal one . . .). Why is someone like Lyell still celebrated today? He is typically lauded in just about every college geology class. It is probably partly due to cultural biases that mythologize certain figures, especially ones who happened to be very rich, and very British, white men. (Granted, it was British white men who were calling Lyell *Cloaca Universalis*.) It's not as if we're even saying Lyell was a bad guy who came up with some great ideas, because he didn't even come up with the ideas in the first place: he stole them from someone else. So why is Lyell lionized? Maybe it was because certain historians of science would rather laud a *Cloaca Universalis* than scientists who were French or Italian and not British: he might have been a universal sewer, but he was our universal sewer. Isn't it time that we stopped foisting a seemingly mythical heroic figure like Charles Lyell on the public, especially when that public includes college students in science classes? It's akin to crediting Sandy Koufax's perfect game to someone else who wasn't even there, yet reported on it as if someone else had pitched those stellar nine innings instead.

Back to James Hall. A meeting with Hall was the first important destination for Lyell after he arrived in the United States. Together they traveled to Niagara to see the falls to gather data on the rate of erosional retreat of its face (Lyell would incorrectly assume it was always slow and gradual and never allowed for occasional catastrophic collapses) and to examine the sequence of the Silurian rocks exposed there. Hall knew these rocks well and shared everything he knew about the rocks and fossils of western New York, showing them off with pride.

Lyell financed his trip largely by singing for his supper: he lectured all over the place, starting with delivering the Lowell Lectures in Boston, which basically paid for his trip. But he kept on lecturing up, down, and around the eastern sea-board. At first Hall and other American geologists freely shared their data. Hall helped Lyell's lecturing gigs with visual aids, and among other things he lent Lyell a geological map that Hall had compiled and drafted but had not yet published. Then there was the rumor that Lyell had made an agreement with Wiley and Put-nam in New York to publish a new edition of *Elements* with "notes and additions" on American geology. Hall worried that it would be viewed as *the* authoritative source on North American geology. Lyell did indeed publish such a tome in 1845, but in London—a two-volume set entitled *Travels in America, with Geological Observations on the United States, Canada and Nova Scotia.* The title contains a weasel word, of course, suggesting to the unwary that the "observations" were all Lyell's.

But perhaps the last straw (again according to Silliman) was a review of "Lyell's works" published in the *New York Tribune* on March 22, 1842, while Lyell was in town to deliver yet more lectures. The reviewer wrote, "We may anticipate the most valuable results from his personal geological exploration of our own country—the prominent points of which he is now spreading before our public in connec-tion with his valuable series of Lectures on Geology."

That did it for Hall. He fired off a letter to the *Boston Daily Advertiser* that was published on March 26. The anti-Lyellian protest letter that Hall wrote and sub-mitted was signed simply "Hamlet."

Why "Hamlet?" No published source known to us explains why. But Niles's lit-erary wife Michelle Eldredge has posed a most satisfactory explanation: in the eponymous play Hamlet's father (also named Hamlet) had been killed by his younger brother Claudius, who thereby ascends to the throne. Young Hamlet's mother, Gertrude, winds up with Claudius, and thus remains queen. This vile usurpation remains unrevealed, generally unacknowledged, and most certainly unavenged. Hamlet the younger contrives to stage a play-within-the-play that reveals the truth to all. The whole thing blows up and everybody dies, with the truth of the usurpation revealed.

We have not found a complete reprinting of Hall's "Hamlet" letter. Silliman does give us a sampling though, from which we extract only the following:

> We wish to see justice done to all; and when editors are to give the idea that
> nothing has heretofore been known of our geology, when is the matter to be
> set right? We might infer that Mr. Lyell has come among a set of barbarians,
> to teach them, for the first time, the name of geology, and that there were

such things as rock in our own country...should he publish a work on American geology, we have no doubt that half the newspapers of the country would hail him as the geologist of America, and give full credit to all that might be said, as if it were the result of his own labors.... We know we are expressing the opinion of more than one American geologist, as well as those engaged in other departments of science, when we raise our voice against this kind of piracy, which we consider as unjust as any other species of robbery.[3]

We Americans was robbed! No one died when Hall's letter was published, though for a time there was a lot of ill will. The influential Yale geologist Benjamin Silliman (who has a college at Yale named after him), possibly an ancestor of the Robert Silliman whose 1995 paper we have been citing, was caught in the political middle between Hall and his many sympathizers among the ranks of young America's most prominent geologists and those who would preserve decorum at all costs and instead rushed to the defense of Lyell. Hall himself went through the usual spectrum of doubts and regrets, on the one hand, and periods of self-justification.

According to Silliman, Hall proposed a book on the geology of North America, a project that never came to fruition. Lyell, on the other hand, who initially suffered consternation when informed of the "Hamlet" letter, seemed mostly concerned with his continuing access to the guidance and information still to be provided by his American hosts, since he had extended his planned stay for another year. A rebuke for past glomming of credit was no biggie for Lyell as long as he still had opportunities to glom credit at some future date. In addition, time healed the wounds, and the matter was largely forgotten. Both Hall and Lyell continued to do their work, though it is interesting that Lyell's rate of scientific publication fell drastically after his 1845 *Travels in America*, with perhaps only the 1863 *Antiquity of Man* of any "novel" consequence thereafter, even if Lyell took flak for alleged plagiarism when that book appeared.

Hall was a thinking empiricist, meaning he stuck to his rocks and fossils but also occasionally proposed original theoretical explanations for geological phenomena. Foremost was his explanation of the subsidence of basins in which, through time, layers of sediments, often chockful of the remains of dead organisms, accumulate. In 1859, he coined the word "geosyncline" for such basins, a word very much still in use in, for instance, Kenneth Hsü's 2004 book *The Physics of Sedimentology*. Alas, we have not been able to find out what Hall thought about

Darwin's ideas on evolution, first published the very same year that he coined the word "geosyncline."

In contrast, Lyell was a keen analyst, synthesizer, and popularizer. In John Clarke's biography of Hall, he remarks of Lyell, "While Lyell was not regarded by his contemporaries as a particularly keen observer in the field, he was by common consent the leader in co-ordinating and philosophic thought."[4] In his 1993 book *The New Catastrophism*, British geologist Derek Ager remarks that the difference between the gradualist, uniformitarian Lyell and the catastrophist Cuvier was that Cuvier actually looked at the rocks. That's quite a heady and insightful remark to come from a British geologist.

Hall expressed some doubt over whether the cultural "imperialism" (he actually used that word) he saw coming from the invading mostly British Europeans was only on the part of some individuals or whether the piratical behavior he worried about and then experienced was the expression of more commonly held national sentiments. The answer to Hall and all others then and now is, of course, that it was and remains both. As Clarke and Silliman make clear, Lyell was not alone in his behavior. But he was the most significant and egregious of the bunch.

Before we leave we must return to Hall, for this is not hagiography, and we in no way wish to replace a British "heroic scientist" with an American one. His are not the shoes that will forever be too big for future generations to fill. The man who was so incensed by Lyell's plagiarism was also legendary for having had his own employees work long hours collecting fossils and then writing about them in works on which Hall would place his, and only his, name. In fact, it was a testament to John M. Clarke's skill and talent that James Hall actually "condescended" to allow him to be included as second author on their aforementioned 1888 work, which Clarke likely almost entirely wrote. Clarke was seemingly grateful enough for this "favor" that he went on to pen his highly complimentary biography of Hall. Maybe Hall felt it was okay to take advantage of colleagues as long as they were Americans? More likely he felt this was one of the perquisites of being the boss. For all that, his memory, and Lyell's, is enshrined in the annals of the history of science.

Before we stray too far from Shakespeare, it bears mentioning that Steve himself was treated as the antagonist, in our view partly unjustly, in a twentieth-century Hamlet affair, but this time the roles of the British and American scientists were reversed. The catalyst was Steve's musings on the Burgess Shale and the nature of contingency in his well-known and well regarded 1989 book *Wonderful Life* (discussed in some detail in chapter 4). *Wonderful Life* was Steve's popular account of the Burgess Shale and what he saw as its key relevance for understanding evolution. He devoted substantial parts of the book to the discovery of this fantastic fossil deposit, which preserves a plethora of denizens of the Cambrian marine realm, including animals not typically preserved in the fossil record because they lack strongly mineralized skeletons. One of these beasts was *Canadaspis perfecta*, which represents an early offshoot of the crustacean evolutionary line (figure 11.3). Understanding of the nature and significance of this extinct marine organism is very much based on the work of our colleague and friend Derek Briggs, G. Evelyn Hutchinson Professor of Earth & Planetary Sciences at Yale University.

In *Wonderful Life* Steve devoted special attention to some of the key scientific figures who discovered and documented the Burgess Shale. One of these, and Steve's principal foil in the book, was Charles Doolittle Walcott, an American paleontologist of great repute in the late nineteenth and early twentieth centuries, who eventually served as secretary of the august Smithsonian Institute. Steve castigated Walcott on several counts, including his distinctly racist ideology and his failure to correctly interpret some of the fossil principles from the Burgess Shale. The heroes of *Wonderful Life*, by contrast, were a set of mid- to late twentieth-century British scientists: the aforementioned Derek Briggs, along with Simon Conway Morris, Richard Fortey, and their thesis advisor, Harry Whittington, all at Cambridge University. Although Steve tried to play the cheerleader for their discoveries, going so far as to suggest that Briggs, Fortey, and Conway Morris deserved bestowal of a Nobel Prize, if such was awarded for paleontology, something about his book definitely rubbed one of the troika, Conway Morris, the wrong way and produced a response worthy of James Hall.

Conway Morris had taken over Whittington's position at Cambridge and may have felt that Gould's portrayal made him seem a little too eccentric. He also was troubled by Steve's take on contingency and the meaning of the Burgess Shale. Conway Morris had adopted a view of evolution seemingly in line with more theistic treatments that saw humans as something special, and if not preordained then at least highly bloody likely. This view was described in his 1998 and 2003 treatments of the topic, *Crucible of Creation* and *Life's Solution: Inevitable Humans in a Lonely Universe*.

What followed the publication of *Wonderful Life* was not a scientific kerfuffle for the ages, as it didn't quite rise to the level of the earlier Hamlet affair, at least

11.3 A specimen of *Canadaspis perfecta* from the Cambrian of Nevada, KUMIP 307021. Photo by Bruce S. Lieberman.

in our view. But it does point out how one needs to choose words very carefully when describing another person's labors, at least while they're still alive. Steve's book undoubtedly made Simon Conway Morris more famous, but whether what happened next was in the category of "no good deed goes unpunished" or "always check with your sources before you publish," or maybe even something else, Conway Morris set out to batter Steve repeatedly in person and in print. (Briggs and Fortey did not fully agree with Steve's interpretation in *Wonderful Life,* but they felt that Conway Morris's subsequent treatment of Steve was little justified.)

Usually in person, when the two interacted and shared the same stage, Steve adopted a "What did I do wrong?" type of posture, especially because he felt that he (unlike Lyell) had given the scientists from across the pond sufficient credit in his writings.

The pièce de résistance of their public interactions came during a notable symposium at Yale University that we regrettably missed, but that thankfully Bruce's wife, Paulyn Cartwright (who is an evolutionary biologist herself), attended as a graduate student at Yale at the time. The symposium, she reported, triggered at least figurative fireworks. At times the air seethed with vituperation between the two, with the metaphorical daggers coming most prominently from Conway Morris's direction. The discussion reached a denouement on the precipice of what today seems befitting of reality TV, when Conway Morris made an off-the-cuff remark about contingency and how if Hitler had fallen out of his pram while an infant then the Holocaust never would have happened. Steve justifiably took offense, as he felt Conway Morris was making light of the Holocaust, and indeed it's never a good idea to invoke Hitler if your subject matter is paleontology (or just about anything else)—see the extensive discussion regarding what is known as Godwin's law.

Neither came to an agreement at the symposium, and their subsequent disagreements became even more strident. Soon after the Hitler comment Conway Morris threw his hands up and left the stage after saying something to the effect of "this is all metaphysical anyway." On a positive note, the exciting aspects of the symposium did at least make paleontology seem like a dynamic rather than a dry and dusty pastime to Cartwright and the other attendees. To that core notion at least we say, "score one for paleontology!" Maybe the nineteenth-century Hamlet affair had done the same for geology in its time.

The interactions between Steve and Conway Morris in print were more placid, at least on Steve's end: he adopted a nonchalant posture, figuring he'd win out in the court of public and scientific opinion in the end. Many pages of text followed that New Haven symposium, including Conway Morris's two books and several of Steve's as well. There was also a famous pairing of side-by-side pieces on the nature and significance of the Burgess Shale by Steve and Conway Morris in *Natural History* magazine, the vehicle of Steve's famous essays. Amusingly, those essays triggered a brilliant letter to the editor in a subsequent issue of *Natural History* from an actor and paleontology enthusiast who remarked that on stage one should always avoid playing opposite Sir Laurence Olivier, and in this debate Steve had played his part.

In retrospect, the interaction between Conway Morris and Steve reminds us not only of the Hamlet affair but also at least a bit of the lead-up to a championship boxing or professional wrestling match. In those types of venues, sometimes vehemently insulting your opponent can pump up viewership and audience interest, and thus benefit the bottom line. Rumor has it that boxers and other types of fighters even conspire to do this type of activity for purposes of enhanced mutual profit. Steve certainly was not conspiring with Conway

Morris, but Conway Morris may have taken a page from many a professional fighter and realized that his appearance in Steve's *Wonderful Life* could well be his ticket to greater fame if only he brought some negative attention or shade to some aspect of Steve's formulation. (Conway Morris was already a well-known and highly regarded scientist, though he lacked the popular audience that Steve had.) Or maybe his motivation was entirely subconscious? The truth is that just as we don't know what makes a scientist famous, we don't know why they do many of the things they do. With luck Conway Morris (or someone else) won't take offense at anything we've written here, though therein could lie a scheme to better publicize this book, redounding positively on net sales. But we've never wanted to enter the ring as prizefighters or professional wrestlers, principally for fear of being pulverized, so we'll leave that type of activity to others better equipped to handle the metaphorical or literal pummeling that inevitably ensues.

When all is said and done, Mark Twain's words seem especially poignant: "If you pick up a starving dog and make him prosperous he will not bite you. This is the principal difference between a dog and a man." Ergo, one needs to be very careful about whom they criticize or praise in print. Sometimes it's better to let sleeping dogs, even well-fed ones, lie, and not name check them until after they've become figurative, or literal, fossils themselves.

CHAPTER 12

When Is a Raptor a Parrot?

The Curious Case of the American Kestrel

W e're paleontologists that study long-extinct marine invertebrates. Thus, if we had to turn our attention to modern life
forms in nature, you might perhaps expect that the things we'd
be most excited to observe would be found by poking around tide pools and
involve searching for starfish or mulling over and under stones in pursuit of mollusks. Undoubtedly, such activities can be pretty cool. But if you asked us what
animal we most enjoy observing in the wild, even in the "wilds" of New York City's
Central Park, hands down we'd say "birds." In their honor, here we tip our cap to
some of the representatives of the "charismatic vertebrate macrofauna," those fascinating, multicolored, beautifully diverse, denizens of the air (usually, anyway):
the Class Aves.

Bird watching is a pastime that many find relaxing and inspiring, even if it
sometimes leaves you with a stiff neck. Still, identifying the precise species you've
seen or heard is not always easy, especially if you're inexperienced or encounter
something novel. In the "old days," and by that we mean the late twentieth or early
twenty-first century, if you saw something new and unfamiliar, absent an experienced bird watcher in your midst, to obtain an identification you'd have one of
two choices. You could try to remember what you saw until you got home, when
you could thereby peruse the titles in your library, or you'd lug around your one
go-to guidebook that you could refer to out in the field. Today's identification
process, however, has been dramatically enabled by electronic resources that simply did not exist just a few years ago. For instance, now you can get an app on
your phone, such as one that we find particularly awesome and efficacious, Cornell University Ornithology Lab's Merlin ID. Fire up the app, and it will ask you
questions about where you saw the bird, what day it was, where it was spotted

(on a bush, on a powerline, and so on); then you get offered a range of sizes, colors, habits, and the like. You choose the correct option for each of those, and then the app provides you with pictures and descriptions of the species it might be, which can be any number of possibilities. You can even take a photo and the app can use that for identification purposes. Cornell also has a great website called ebird.com.

Circling back to our discussions of punctuated equilibria and stasis, we find it amazing that anyone would challenge the notions that stasis prevails and, further, that species are real when some bird, let's say the Eastern phoebe, always shows the same diagnostic characteristics of color, shape, and size, and warbles the same distinctive call. Indeed, this is one of the things that we believe makes bird watching so exciting and entrancing. There is a profuse diversity of traits and appearances that exist among the various species that can be encountered, yet these traits show consistency within certain narrowly circumscribed groups: species.

Also awesome, these birds display consistent behaviors (beyond just the behaviors involving their songs and calls). For instance, the Eastern phoebe is known for its "tail wagging." This is one of the omnipresent behaviors that eases identification of the species. Whether in Florida or the Northwest Territories of Canada, when you see an Eastern phoebe, there's going to be a darn good chance it's bobbing its tail up and down in a distinctive way. Score one more for stasis, the reality and individuality of species, and hence macroevolution.

Electronic resources have not only stepped up our ability to identify birds, but they've also made it easier to see where species occur throughout the year. Again, the folks at the Cornell Ornithology Lab are doing amazing things with range maps of species, showing how they change dynamically week by week across the different seasons. It shows us, for one thing, that you're unlikely to spy an Eastern phoebe in the northern or central part of the United States in the month of November. These maps are built up from observations by amateur and professional bird watchers.

Perspicacious bird watchers want to know that their identifications are accurate. They also want to know more about how the myriad bird species are classified, which is a question of bird phylogeny or evolutionary relationships. Coincident with the recent advances in technology assisting bird watching there have been new developments in bird phylogeny, and it is to these that we will turn as they provide a fascinating perspective on the affinities of another bird.

THE AMERICAN KESTREL

One November Bruce pulled up to a traffic light near the banks of the Kansas River, or Kaw to the locals, a relatively large river by Midwestern standards that serves as a corridor for migrating birds. In Bruce's peripheral vision a robin-sized bird alighted on a streetlamp. It was evening, but in the gloaming all Bruce could discern was that it wasn't a robin; its breast had a slightly reddish tinge above white, its head and back were dark, maybe bluish-gray, and it had a beak distinctly larger than the type common to flycatchers or warblers. The light changed, and although in Kansas, as against New York City or Boston, a brief pause after a green signal is considered permissible and does not precipitate attendant honking, there clearly was no time to fire up Merlin.

There was a vague sense that the size, robust beak, reddish breast, and blue-gray back and tail possibly connoted a male rose-breasted grosbeak, but Bruce considered the hypothesis only tentative, because, although technically not colorblind, Bruce has poor color judgment (and fashion sense). Given the muted shades brought on by twilight, Bruce's initial assertion of "reddish" when it came to the bird's upper breast might actually constitute a much broader palate of the visible spectrum of light than typically consigned to the color red.

After he arrived at his home a short time later, subsequent research made it clear to Bruce that he had not seen a grosbeak, whose breast was way too red; its tail was also shorter and stubbier than the bird Bruce had seen. Moreover, Cornell's ebird.com revealed that rose-breasted grosbeaks are not likely to be found in Kansas in November: only a single November sighting had been made in Lawrence in the preceding ten years. The initial hypothesis was not looking good, and there was more exploring to be done.

We both have been blessed with careers at natural history museums. Such employment has many perquisites, not the least of which are the opportunities to interact with expert naturalists. One, pivotal to this story, is Mark Robbins, the ornithology collections manager at the University of Kansas (KU) Biodiversity Institute. Thanks to him the hypothesis was revised: this was afforded by his knowledge and the opportunity Mark provided to examine actual specimens. In other chapters we have sung the praises of natural history museums and the treasure trove of biodiversity specimens they contain, for in this case they provide

12.1 Specimens of the American Kestrel in the KU Biodiversity Institute. Photo by Mark Robbins.

fundamental data about bird morphology, distribution, and habitat. Bruce's bird was actually a male American kestrel (figure 12.1).

These birds are orange, not red, but we suppose orange can look red at night, especially if you're myopic. These tiny birds of prey love to perch on lampposts at dusk to be on the lookout for food. Furthermore, November is also a good time to be an American kestrel in Lawrence, Kansas, according to the folks at Cornell.

Not only was Bruce to be disabused of his initially faulty identification, but we were also flummoxed by our lack of knowledge about birds of prey and their origins and affinities. Some might associate this bird, in an evolutionary sense, with hawks, and many other hawks are common in this part of the world, including the red-tailed hawk. But kestrels aren't hawks, they're falcons. Same difference, or close enough, right? Wrong. Big time. Again, thanks to Mark Robbins, it became clear to us that kestrels and falcons are not even closely related to hawks. An outstanding study by Alexander Suh, now at the University of East Anglia in the United Kingdom, and colleagues, published in the journal *Nature*

Communications in 2011, and based on an analysis of bird genomes shows why. The result by Suh and colleagues bolsters previous research, including work by Shannon Hackett and colleagues published earlier in the journal *Science* in 2008.[1]

Suh and Hackett and their colleagues showed that kestrels and falcons are in fact closely related to parrots. By contrast, hawks and eagles sit well outside of these and nest, evolutionarily, with Old World vultures. Macroevolution is like family: you don't always get to pick who you are related to. You may love all your relatives, and if so great, but if not, tough luck. At the scale of the history of life, if you're a kestrel you get cousins that are brightly colored, long-lived, and can say things like "pieces of eight." If you're a hawk, your cousin is a carrion eater with a proclivity for munching on roadkill.

We can be awed by the similarities between kestrels and hawks, and the thought that these seem to have evolved independently (the biological term typically invoked is "convergently"). But we shouldn't be overawed. At the scale of the history of life, sometimes convergence happens. Moreover, kestrels were at least partly grouped with hawks based on weak evidence: traits such as "likes to eat meat," "flies fast," "reddish," and so forth. Detailed comparisons not only of molecules but perhaps also of morphology, including that somewhat stubby beak we mentioned previously, might bolster the kestrel/parrot affinity.

To get here we've utilized information based on the latest technology, like dynamic range maps and studies that incorporate large portions of the genome to reconstruct evolution. But just as important, this tale was driven by expert knowledge and the ability to access specimens collected decades ago.

MUSINGS ON SPANDRELS

As we said, some of the traits shared by kestrels and hawks appear to be convergences that evolved independently. It is very likely that these would have evolved because certain features associated with a predatory lifestyle in birds were favored by natural selection. But what about all those other traits that kestrels and hawks share? These in fact comprise most traits, and one might wonder how scientists view these. Throughout the second half of the twentieth century, if one had sampled most evolutionary biologists the response one would have gotten was that of course these were adaptations too. Thankfully, in the interests of accuracy and clarity this is no longer the case. And a big part of the reason for that shift has to do with Stephen Jay Gould and Elisabeth Vrba. Let's start with Steve, proceeding on a chronological basis.

In 1979, Steve, along with Richard Lewontin, whom we mentioned in chapter 6, published a paper with the ostentatious, yet also superb title: "The Spandrels of San Marco and the Panglossian Paradigm: A Critique of the Adaptationist Programme."[2] It begins: "An adaptationist programme has dominated evolutionary thought in England and the United States during the past 40 years. It is based on faith in the power of natural selection." Perhaps never before had someone had the sheer audacity to title a scientific paper after a fifteenth-century Venetian cathedral while also name-checking a character featured in the works of the eighteenth- century philosopher Voltaire. Mad props and kudos to them! Even if they also came across as a little pedantic. Cited by more than 10,990 other scientific articles as of February 3, 2024, a prodigious figure, the paper went well beyond a snazzy title in its significance. It identified a key problem in the field of evolutionary biology: that there was an inherent bias to presume that every trait of organisms was perfectly adapted to perform a specific function. Steve and Richard Lewontin partly reached this conclusion as a deduction arising from the existence of punctuated equilibria, or punk eek (see chapter 2). The 1972 punk eek paper by Niles and Steve had, among other things, outlined that there was a bias toward gradualist thinking in evolution while demonstrating that species were not continually changing and progressing throughout their existence. This meant that the traits of organisms within those species were not continually adapting as well.

Steve and Richard used the example of the small triangular spaces in the St. Mark's Cathedral, called spandrels, that arose as the space created when an architect built a dome in the Middle Ages. Back then, before prefab steel, lofty domes needed to be supported by stately arches. Where the arches supporting the dome met there would be a triangular space. The architects and designers of the cathedral beautifully decorated and painted these spandrels, such that they blended in perfectly with the religious allegory the edifice conveyed. But Steve and Richard's key point was that the spandrels weren't put there to convey that allegory; instead, they simply had to exist, like the statistical laws of science. Given that constraint, the architects and designers took advantage of them. Maybe most biological traits were like that as well: existing for another reason or as a constraint, with nature subsequently taking advantage of the opportunity.

Steve and Richard were not the first scientists to make this point. Some of those that challenged Darwin's ideas shortly after they were first published argued that Darwin had over emphasized the notion of traits being shaped by selection. But Steve and Richard certainly raised the most substantive and effective challenge in the second half of the twentieth century. Intriguingly, there are precursors to aspects of Steve and Richard's arguments in one of the works of George Williams, which is partly ironic given that he is sometimes viewed as among the most hyperadaptationist of adaptationists. Yet at least early in his career this was not entirely true. For instance, in his classic 1966 treatise *Adaptation and Natural Selection* he considered the intriguing case of flying fish. Do biologists need to posit any special adaptations to explain how the fish *returns* from the air to the water after leaving it? Some were trying to do just that. Here is what Williams wrote:

> It is clear that there is a physiological necessity for it to return to the water very soon; it cannot long survive in air. It is, moreover, a matter of common observation that an aerial glide normally terminates with a return to the sea. Is this the result of a mechanism for getting the fish back into water? Certainly not; we need not invoke the principle of adaptation here. The purely physical principle of gravitation adequately explains why the fish, having gone up, eventually comes down.... In this example it would be absurd to recognize an adaptation to achieve the mechanically inevitable.[3]

It was a highly cogent rejoinder against the tendency to posit a purpose for everything, which was in fact the practice of the notorious but hypothetical Dr. Pangloss whom Steve and Richard referred to in their paper.

Regrettably, Williams himself later in his career, along with some of his best-known followers, including Richard Dawkins, did not always heed this rejoinder, or at least seemed to think it did not apply to their own speculations, thereby proving one of the things Steve and Richard had been all about in the first place. In Williams's case this especially involved his ideas on Darwinian medicine.[4] For those of you unfamiliar with it, Darwinian medicine seems to have been, while not a creation of insurance companies, then a grand excuse for them to deny coverage for treatments of various ailments running the gamut from fevers all the way up to impending death. The argument invoked was that our bodies are so finely tuned and adapted that any treatment we apply is likely to interfere with the body's mechanism for curing whatever ails it. So, for instance, it was asserted, don't take ibuprofen in the event of a fever, for it interferes with the evolved immune response to kill infectious agents.

Williams early on was at the forefront of the Darwinian medicine movement, which in our view is mostly wrongheaded. Seemingly he had forgotten about antibiotics and how they have saved hundreds of millions of people from strep throat turning into scarlet fever, or vaccinations, which, in the case of just one instance, tetanus, prevented more moderate wounds from morphing into pernicious and deadly lockjaw. Human beings, if so perfectly honed and shaped by natural selection, surely would have had no need for such antibiotics or vaccinations. But this transformation occurred late in Williams's career, and thus perhaps he can be at least somewhat excused.

EXAPTATIONS

One of the most exciting ideas that came out of Steve and Richard's paper was another paper published three years later by Steve and Elisabeth Vrba, "Exaptation—a Missing Term in the Science of Form," which introduced that term. Exaptations were something hinted at in Steve and Richard's paper, adumbrated in a scant paragraph where they were not fully defined. Elisabeth right away saw the broader significance of this paragraph that the authors had left to tidy up later (or perhaps never). Elisabeth, given her interest not only in macroevolution but also in reconstructing phylogeny (evolutionary history), realized that as transformations occurred traits that were originally under natural selection for one function might later come to be selected for another function. Or they might evolve neutrally, but then later come to be coopted. Simply labeling these traits "adaptations" would miss key aspects of what was going on.

The paper provided several examples of the phenomena, one of which beautifully illustrated how paleontology can at times be a predictive science. They posited, "If we ever find a small running dinosaur, ancestral to birds and clothed with feathers, we will know that early feathers were exaptations, not adaptations, for flight."[5] The function of feathers and what they were selected for when they first evolved instead might have been for insulation and regulation of body temperature. Only later, and with some subsequent modification, would feathers be able to generate the lift needed to function in flight.

In 1982 Elisabeth and Steve's statement was remarkably prescient. Today, feathered dinosaurs are known and beloved by adults and children around the world; they have graced postage stamps and covered top-selling magazines. At least as a child, even that most formidable of carnivorous dinosaurs, *Tyrannosaurus rex*, was festooned with feathers. But their existence would not be confirmed for more than fifteen years, when they were recovered in spectacular paleontological deposits from China that preserved skin impressions.

Elisabeth brought this idea to Steve, convinced him that it was significant and deserving of extra discussion beyond what Steve and Richard had parlayed, and hence was born their first scientific collaboration and coauthored paper. The drafting of this paper was absolutely a trying time for Steve, as it occurred while he was receiving chemotherapy for mesothelioma. His weight had dropped precipitously, he was very ill, and Elisabeth reported that he was frequently running midsentence to the restroom due to bouts of nausea. The sheer persistence and gumption that he displayed during the work was inspiring. The paper was ultimately published in *Paleobiology*, the journal that was born out of the book that produced the original punk eek paper in 1972, a fitting home indeed. Although not quite as oft cited as the spandrels paper, it still clocks in with a hefty 6,363, and climbing, citing publications as of February 3, 2024. It is known as Gould and Vrba's 1982 exaptation paper, and at the bottom of the first page it reads, "An equal time production; order of authorship was determined by a transoceanic coin flip."

As writers we are always interested in teachable moments, and for you budding scientists out there, something as potentially contentious as order of authorship should not be left up to chance. Not that it ultimately matters too much. In our experience scientists rarely read papers, they even more rarely cite them, and they hardly ever understand them. Indeed, we've both written papers that say something to the effect of "this concept isn't really valid." Later we get a notification from Google that says your paper on this has been cited by article *X* written by author *Y*, which at first gives us a warm fuzzy feeling until we go check out that citing paper and find that we've been cited in support of the very concept that we said in the paper was likely hogwash. That's academia for you: sometimes you just have to take what you get.

Meanwhile, we're hoping, given its evolutionary roots, for a talking kestrel to turn up.

CHAPTER 13

What's Your Favorite Trilobite?

Walter Winchell Wouldn't Have Cared

One of the quintessential aspects of being a graduate student in the Department of Geological Sciences at Columbia University in the 1980s and 1990s, Bruce's vocation at the time, was taking the dreaded "Shuttle to Lamont." Although there were positive aspects of the program, a distinctively negative aspect was its fragmented nature: the department comprised several smaller campuses scattered over parts of New York City and the larger metropolitan area. On any given day meetings, classes, seminars, and social gatherings could be held at any of these sites, meaning that a frenzied commute might be in order. For a student to travel between two of the major locales, Columbia's Morningside Heights campus and what was then the Lamont-Doherty Geological Observatory (today's Lamont-Doherty Earth Observatory), the Shuttle to Lamont was the sole means of transport. The shuttle, really a poorly appointed minivan, ran every few hours from about 9:00 a.m. until about 3:30 p.m. from a spot in the vicinity of 120th Street and Amsterdam Avenue, hard by Columbia's Morningside Heights Campus, up to the location of Lamont in the "hamlet" of Palisades on the western bank of the Hudson River.

Bruce is going to assert that graduate school is better without having the added burden of mad dashes to a suburban research observatory that required a frantic ride aboard the #1 subway train, necessitated by the fact that Bruce's and Niles's office was in the august American Museum of Natural History. So, if Bruce was conducting research during those grad school years, which for better or worse he almost always was wont to do, he would be at the AMNH. The quickest way to make it from the AMNH to the Morningside Heights campus, and the shuttle stop, was via the #1 train (proceeding from the 79th Street to the 116th Street stations). The mad dashes occurred because Bruce dreaded the commute to Lamont, so he left departing the AMNH until the last possible second, which ironically ensured that the commute was all the worse, and he dreaded even more missing the last shuttle to Lamont. In retrospect, at least, Bruce can acknowledge that he survived it.

There happened to be several distinctive things associated with the Columbia geology grad school experience, but the one Bruce will single out here involved the fact that there were two drivers of the Lamont shuttle, one named "Walter," the other "Winchell." Seriously, what are the odds of this curiosity? To Bruce it seemed inimical to the notion of a disordered universe that a thirty-minute ride, including stretches on the West Side Highway, the George Washington Bridge, and the Palisades Parkway, could partly resuscitate a journalist consigned to the dustbin of history, Walter Winchell, who dominated the airwaves from the 1930s to the 1950s as America's preeminent media personality, with an outsized reputation and a prodigious reach attained via newsprint and radio, with a reported readership of between twenty million and fifty million daily. Winchell was further known for his distinctive tagline: "Good evening, Mr. and Mrs. America from border to border and coast to coast and all the ships at sea."

The bulk of his career was bracketed by an early push to battle Nazi appeasement and a late push to appease Joseph McCarthy. A somewhat notorious episode in Winchell's life occurred in 1939, when he managed to get feared gangster and cofounder of the crime syndicate known as Murder Incorporated, Louis "Lepke" Buchalter, to turn himself in to the feds after he had gone on the lam to escape Thomas E. Dewey's prosecutorial wrath. Lepke had avoided the law for more than two years. Rumor has it that courtesy of "Hollywood Godfather" Gianni Russo, Buchalter spent part of his time hiding out in one of the towers of the Brooklyn Bridge in a tiny makeshift "apartment" complete with radio. Lepke was ultimately coaxed out of hiding via Winchell's radio transmitted promise, reproduced in the *New York Daily News*, that "I'll tell John Edgar Hoover. . . . I'm sure he will see to it that Lepke receives his constitutional rights and nobody will cross him."

One thing is made clear by this narrative: the way people guaranteed someone's safety in the old days was quite fluid. Still, the more things change, the more they stay the same, as such guarantees meant as much then as they mean today: bupkes, zilch, nothing. Lepke was eventually fried in Sing Sing prison's electric chair, "Old Sparky," albeit several years later and after all his appeals had been exhausted, the only major U.S. mob boss to receive the death penalty. Winchell's personal guarantees for constitutionality had either been left by the wayside or he did not consider electrocution anathema to the Constitution's Eight Amendment prohibiting cruel and unusual punishment.

This seeming betrayal is one, albeit certainly not the greatest, of reasons why Winchell has been described as "odious" and "loathsome" by various sources, including the *New York Times*. Indeed, it is perhaps a dubious ethical question as to whether Lepke, who may have engineered two hundred murders, deserved a fair shake. Yet if you're going to lie to the capo of Murder Incorporated, who wouldn't you lie to?

It is impossible to tease apart Walter Winchell's take on different aspects of paleontological research. Practitioners of *Citizen Kane*-era journalism were curiously shy when it came to going on the record about the causes and consequences of the Late Cretaceous mass extinction, and Winchell was no exception. Yet his shared connection with the eponymous drivers of the Lamont minivan— all three were also former residents of New York City—may provide some clues. And it was either Walter or Winchell who remarked after querying Bruce's friend and Niles's former student and now highly accomplished scientist Greg Edgecombe about the nature of his PhD studies, centering on a charismatic fossil arthropod group: "Trilolights? Why the hell are you wasting your time studying trilolights? You should be out . . ." (fill in with any activity you think Columbia University shuttle drivers are likely to find more appealing).

So, while it is true that if you asked most people the scientific name of their favorite trilobite you'd probably get a blank stare, it may be surmised that Walter Winchell would not have cared one whit and may well have provided a snarky riposte. But trilobites are central to the geological enterprise. This archaic group, once possibly the dominant animal life form, has for the last 250 million years been extinct. Such is the capricious nature of evolution and existence: one minute you're on top, the next you're gone; sometimes you're the windshield, sometimes the bug.

Despite their no longer halcyon status, or perhaps because of it, trilobites continue to inspire some for their relevance to understanding evolution and some for their sheer aesthetics, and some for both qualities. We are proud to classify ourselves in the latter category. What about the expiration of Walter Winchell's halcyon status? It seems to be partly related to the fact that he had a face for radio and thus mostly vanished from the spotlight after television became the media of choice. However, his embrace of Joseph McCarthy was dastardly, and a downfall precipitated by that would certainly have been karmic.

The Cambrian period (specifically, between about 525 million and 485 million years ago) represented the salad days, to evoke Shakespeare's *Julius Caesar*, of the lovely trilobite, and thus it may be only fitting that Bruce's favorite trilobite is a Cambrian "bug," the ~505-million-year-old *Olenellus getzi* (figure 13.1) known from the Kinzers Shale of Lancaster County, eastern Pennsylvania. The holotype, or defining specimen, of this species is spectacular, residing in the collections of the outstanding Yale University Peabody Museum of Natural History.

Why is it Bruce's favorite? He suspects the reasons are manifold. For instance, he first encountered this species and the paradigm specimen that represents it as the type during his own salad days, when he was a postdoc at Yale. He was young, he was a newlywed, and he was getting paid to do research. Although given that he was living in New Haven, Connecticut, he should perhaps instead refer to these as his "pizza days," since one of the best things to spend a stipend on was New Haven pizza, which Bruce, and many others, consider the world's greatest. Bruce had very few other obligations beyond the pizza—a choice and not an obligation—aside from paleontology and macroevolution.

Carl Dunbar, the taxonomic author of *Olenellus getzi*, was a paleontologist with a Kansas and Yale connection, and thus Bruce always felt some kinship to him. Carl was born and raised in Hallowell, Kansas, a place whose precise location merited little interest until we found out that it is a mere ten miles away from the site of the world's second largest steam shovel, Big Brutus, once adept at ripping apart and despoiling the earth in search of coal and today basically a giant tchotchke that sits in a field waiting for the occasional tourist to drive by and snap a picture.

Dunbar had a premonition that he would end up a grower and harvester of wheat, and there would have been nothing wrong with that. Wheat farmers are important; wheat is, after all, a key ingredient of pizza. But after high school he decided to go to Lawrence in search of a degree at the University of Kansas. To borrow from Robert Frost, "that made all the difference" for Dunbar. As far as he knew, upon graduation Carl was going to return to Hallowell and become a wheat farmer of local repute. However, the vicissitudes of human life in microcosm are documented in Dunbar's memorial biography, available online and open access as a PDF from the Geological Society of America, penned by the late Karl

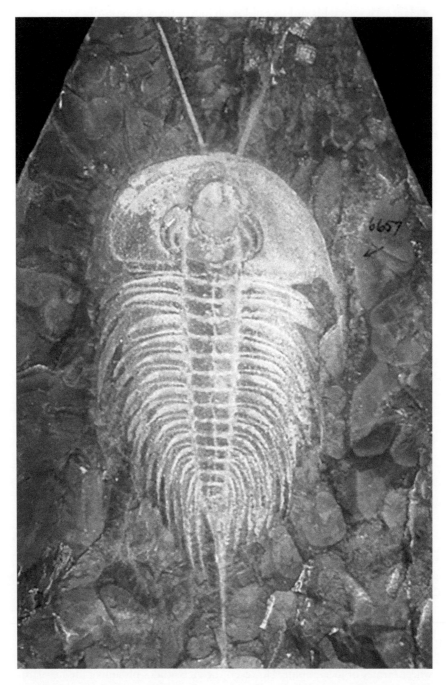

13.1 The holotype specimen (YPM 6657) of *Olenellus getzi* from the Yale University Peabody Museum of Natural History. Photo by Bruce S. Lieberman, used with permission, courtesy of Tim White, Derek Briggs, and Susan Butts.

Waage, himself a former professor of paleontology at Yale University. In the biography, Waage writes that a "visiting uncle asked whether (Dunbar) would study any geology. 'What's geology? I never heard of it,' replied Carl. The uncle confessed that he didn't know much about it himself, but he had a friend who was a geologist, and he thought it was great. According to Carl, 'Three days later, I was enrolled in a geology course, and the instructor was W. H. Twenhofel.'" Twenhofel in turn was a Yale alum, Dunbar became enamored with paleontology thanks to him, and after graduating from Kansas he went on to receive his doctorate at Yale.

The rest, as they say, is history. Along the way Dunbar eventually became a professor at Yale, thanks to the support of his doctoral advisor, Charles Schuchert, himself an eminent paleontologist. This was a common occurrence in the old days, when nepotism prevailed, although Dunbar was an excellent candidate and there may have been few other competitors for the position in those days. Stephen Jay Gould himself benefited from this very same type of system. As he was finishing up his PhD at Columbia, Steve's advisor, Norman Newell (discussed in detail in chapter 7), got a call from Bernie Kummel, on the faculty at Harvard and himself a Newell student. Bernie inquired whether Norman had anyone he could recommend for a newly opened position. Norman said, "We've got a good one here," and before too long Steve was moving on up to Cambridge with an assistant professorship in hand. (He fully deserved the position, and we can think of no one better for it.) Niles may have benefited from a similar type of circumstance, for he came to work as a curator beside Newell in the very department he had been trained. Bruce doesn't think he was necessarily the recipient of this sort of largesse, at least in the blanket sense, having interviewed for and been rejected by a host of institutions great and not so great. Yet it may be no coincidence that he ended up with a gig at Kansas, where Newell and Dunbar had been all those years ago.

During his tenure at Yale, Dunbar ultimately became director of the Yale Peabody Museum of Natural History and played a pivotal role in causing Rudolf Zallinger's spectacular mural *Age of Reptiles* (figure 13.2) to become a centerpiece of the Peabody's exhibits. Along the way Dunbar won numerous awards, including election to the National Academy of Sciences, and he was even an observer at the hydrogen bomb tests conducted at Bikini Atoll in 1946. A man with a "pleasant, business-like demeanor," further called by Waage "friendly but not demonstrative" and "disliking pomposity and self-aggrandizement," he died at the age of eighty-eight. We have been unable to determine his opinions on the relative merits of New Haven pizza.

13.2 Portions of Rudolf Zallinger's "Age of Reptiles," with a sampling of dinosaurs in the foreground, from the Yale University Peabody Museum of Natural History's Great Hall of Dinosaurs. Photo by Bruce S. Lieberman.

Back to *Olenellus getzi*. The defining type specimen of the species is impressive (as are most complete specimens), spanning more than 20 centimeters in length, not counting the spine (more on that in just a bit), with a bright orange color imparted from the iron-rich minerals abundant in the local rocks. Notably, there are two prominent antennae emerging from underneath the front of the head shield; these are seldom found in *Olenellus getzi* specimens (the type specimen represents one of the few and best examples known) and indeed are decidedly uncommon for trilobites in general: of the roughly twenty thousand species of trilobites, most are known solely from their strongly mineralized external skeleton, while evidence of their soft tissues, including appendages like antennae, is rare and found from perhaps fifteen species. This disparity in preservation is typical of the entire fossil record: soft tissues are by their nature only rarely fossilized. However, the Kinzers Shale is well known for preserving soft tissue, exemplified by numerous specimens belonging to possible crustacean ancestors.

The holotype of *Olenellus getzi*, along with numerous other specimens, was collected by Noah L. Getz and his son from a small stone quarry on the Getz family farm a mile north of Rohrerstown and two and a half miles west of Lancaster off the Harrisburg Pike. Today the site, now inaccessible, is close to a Baptist church, and it appears to sit on land lapped to the south by a sea of subdivisions. In bygone days the view would have been far more bucolic.

The history of the Getz family in the region is almost as intriguing to trace as the taxonomic history of the trilobite, and indeed it is far more extensive, given that Dunbar's publication on the species dates to 1925 and it has been relatively little considered since. According to *The Biographical Annals of Lancaster County*, the first family member to arrive in America was John Jacob Getz, "who came to America in the good ship '*Dolphin*' and landed in Philadelphia in 1738, his home having been in Pfalz, Germany. After a short residence in this country, he went back to Germany, but finally returned and located in Lancaster County, settling on Chestnut Hill and occupying a large extent of country, some 400 or 500 acres of the choicest land of this fertile county." The work continues to Noah L. Getz, who, with his wife, "settled on a tract of 120 acres of land which he had purchased from his father; and upon this place he has made his home, and added many valuable improvements, in 1879 erecting tobacco sheds and several smaller buildings." According to the Yale Peabody Museum these tobacco sheds and other smaller buildings may have been built of the local fossiliferous rocks that comprised the source of *Olenellus getzi*, which, as an amateur collector, Getz recognized as something different. Amateur fossil collectors have done so much to benefit the science of paleontology, and Dunbar bestowed Getz's name on the magnificent collection, including YPM specimen #6657, which the Getz family donated to Yale. Dunbar remarked in his 1925 *American Journal of Science* publication, where he described *Olenellus getzi*, that the collection "is a monument to fine enthusiasm, and one which shows what indispensable aid the labors of laymen may yield to the cause of science."

Another distinctive aspect of this once hale and hearty trilobite is the spine that appears to protrude from its back end. This spine was a red herring that led a German scientist named K.-E. Lauterbach on a brave but ultimately foolhardy errand in the early 1980s. He used the spine to posit in various papers that this particular species of trilobites and its close kin were not actually trilobites at all but instead more closely related to chelicerates, the arthropod group including spiders. This potentially substantial shift in the classification of trilobites of the genus *Olenellus* was performed via Lauterbach's contention that the spine was an evolutionary equivalent or homolog to the terminal spine or telson of horseshoe crabs (figure 13.3).

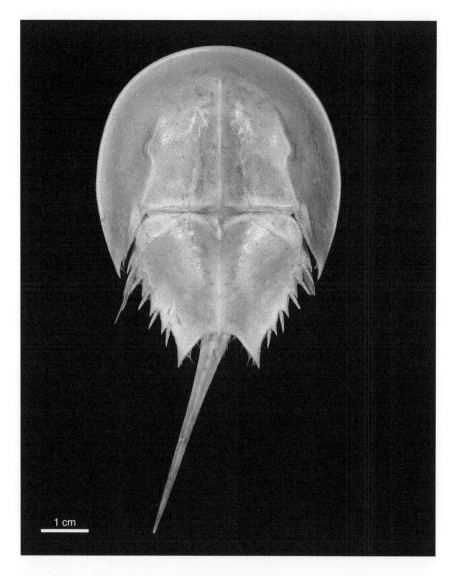

1 cm

13.3 The horseshoe crab *Limulus polyphemus* from the East Coast of the United States, KUMIP 405574. Photo by Natalia López Carranza (KU).

Alas for Lauterbach, or at least his hypothesis, the spine is not a terminal spine or telson projecting from the posterior end of the body at all. Instead, it is a spine that would have projected up in the air; further, it lies not at the very end of the body, but rather there are some trailing segments that culminate in a tiny tail that is often covered by the larger spine when fossilized.

No, *Olenellus* and its relatives are bona fide trilobites, the most ancient and basal members of the group, a relationship Greg Edgecombe and others have documented. Where ultimately Lauterbach went wrong with these spines is that he did the evolutionary equivalent of treating butterfly wings, pterosaur wings, bird wings, and bat wings as homologous, supposing that they embodied the same structure in different organisms.

However, we should not castigate Lauterbach for his faux pas. It is not always easy to forge evolutionary connections between modern and ancient creatures. Sometimes we try and sometimes we fail. At least Lauterbach tried. And here's to those who try: Carl Dunbar, Noah Getz, Walter, and Winchell. They all helped make the world a better place, even if they didn't all care about trilobites (or trilolights).

Niles's favorite trilobite is of course *Eldredgeops rana* (figure 2.1), and the reason for that should be obvious. It's what led to his discovery of punk eek. And we think for that same reason it would be Steve's favorite trilobite too. (Niles has detailed the discovery of punk eek and the nature of *Eldredgeops rana* elsewhere, especially in the book *Time Frames*, so we hope the interested reader will check that out as well.) Granted, Steve liked snails a lot more than trilobites. And he had the wit to pick fossils snails from Bermuda to be his "go to gastropods." Fieldwork in Pennsylvania Dutch country can be downright pleasant, but Bermuda likely has it beat in most people's books. Steve would marvel at the geometry, imbrication, and ornamentation of snail whorls, among other features, and many of these traits he would dutifully note and strive to quantify for analysis in his taxonomic treatments. These were more extensive and erudite than most folks gave him credit for. Regrettably, Walter, Winchell, or Walter Winchell never asked him, "Why the hell are you wasting your time studying those yucky snails?" If they had, we believe he would have said, "Because they inspire me." And then he likely would have gone on to provide a lengthy excursus about how snails carry their entire life spans on their back, have an exceptional fossil record, are an exceptional system to study evolution, and much more, that extended long past the point when the van had been put into drive, the driver had merged into Amsterdam Avenue traffic, and the friendly inquisitor had tuned out.

He was always loquacious when it came to his passions: snails, baseball, the operettas of Gilbert and Sullivan, evolution. Steve's life and work continues to inspire us, long after he departed for that great proverbial paleontological hunting ground in the sky. Yes, his work goes in and out of fashion, for as Rudyard

Kipling remarked in *The Man Who Would Be King*, "the wheel of the world swings through the same phases again and again. Summer passed and winter thereafter, and came and passed again." But even when it's winter somewhere, it's summer somewhere else. Although this story is over, Steve's lives on. May his legacy only continue to spread and grow down the seasons.

Acknowledgments

We extend our special thanks and extreme gratitude to Miranda Martin and Brian Smith of Columbia University Press for all their editorial and other assistance throughout the process of turning our manuscript into a finished book: it would not have been possible without their support. We very much appreciate the help of Natalia López Carranza, Biodiversity Institute, University of Kansas, who took several photographs used here. Thanks also go to Malcolm Baker, Brendan Bradley, Derek Briggs, Alexandra Briseno, Susan Butts, Jonathan Hendricks, Kathleen Huber, Teresa Manera, Barbara Murphy, Gillian Newell, Tim Noakes, Richard Prum, Mark Robbins, Chris Schell, Ilya Tëmkin, Brad Utter, Bridgett vonHoldt, George Vrba, Jonathan Way, and Tim White, who generously contributed figures and/or provided permission to use them. Greg Raml graciously searched within and shared materials from the Gottesman Research Library of the American Museum of Natural History. Bruce acknowledges financial support from the National Science Foundation and Biodiversity on a Changing Planet award 2225011. We also very much appreciate Gregory McNamee's skillful copyediting, Kathryn Jorge and Zachary Friedman's assistance with various editorial matters, and the thoughtful comments and suggestions of three anonymous reviewers. Last, we thank Paulyn Cartwright and Michelle Eldredge for their helpful insights regarding the text.

Notes

2. ASLEEP AT THE SWITCH

1. S. J. Gould, *Ontogeny and Phylogeny* (Cambridge, MA: Harvard University Press, 1977).
2. J. N. Thompson, *Relentless Evolution* (Chicago: University of Chicago Press, 2013).

3. SURVIVAL OF THE LAZIEST

1. G. G. Simpson, *Tempo and Mode in Evolution* (New York: Columbia University Press, 1944).
2. L. C. Strotz, E. E. Saupe, J. Kimmig, and B. S. Lieberman, "Metabolic Rates, Climate and Macroevolution: A Case Study Using Neogene Molluscs," *Proceedings of the Royal Society, Series B* 285, 20181292 (2018): 1-6.
3. S. J. Gould, *The Structure of Evolutionary Theory* (Cambridge, MA: Harvard University Press, 2002).

4. TIME'S ARROW, TIME'S CYCLE, TIMES SQUARE

1. S. J. Gould, "Eternal Metaphors of Palaeontology," in *Patterns of Evolution*, ed. Anthony Hallam (Amsterdam: Elsevier, 1977), 4.
2. D. Sepkoski, *Rereading the Fossil Record: The Growth of Paleobiology as an Evolutionary Discipline* (Chicago: University of Chicago Press, 2012).
3. See D. M. Raup, S. J. Gould, T. J. M. Schopf, and D. S. Simberloff, "Stochastic Models of Phylogeny and the Evolution of Diversity," *Journal of Geology* 81 (1973): 525-42; D. M. Raup and S. J. Gould, "Stochastic Simulation and Evolution of Morphology: Towards a Nomothetic Paleontology," *Systematic Biology* 23 (1974): 305-22; S. J. Gould, D. M. Raup, J. J. Sepkoski Jr., T. J. M. Schopf, and D. S. Simberloff, "The Shape of Evolution: A Comparison of Real and Random Clades," *Paleobiology* 3 (1977): 23-40.
4. S. J. Gould, "The Promise of Paleobiology as a Nomothetic, Evolutionary Discipline," *Paleobiology* 6: (1980): 96-118.
5. The formal title of Windelband's 1894 "Rectorial Address," as translated into English, is "History and Natural Science," *History & Theory* 19 (1998): 169-85.

6.   A. Silverstein, "An Aristotelian Resolution of the Idiographic Versus Nomothetic Tension," *American Psychologist* 43 (1988): 425–30.

7.   S. J. Gould, "The Molluscan Fauna of an Unusual Bermudian Pond: A Natural Experiment in Form and Composition," *Breviora* 308 (1968): 1.

8.   S. J. Gould, N. L. Gilinsky, and R. Z. German, "Asymmetry of Lineages and the Direction of Evolutionary Time." *Science* 236 (1987): 1437.

9.   N. Eldredge, "Differential Evolutionary Rates," *Paleobiology* 2 (1976): 174–77.

10.   S. J. Gould, *Time's Arrow: Time's Cycle* (Cambridge, MA: Harvard University Press, 1987), 178.

11.   Gould, *Times's Arrow*, 97.

12.   Gould, *Time's Arrow*, 196.

13.   S. J. Gould, *Wonderful Life* (New York: Norton, 1989), 4.

14.   Gould, *Wonderful Life*, 274.

15.   S. J. Gould, "Introduction: The Coherence of History," in *Early Life on Earth: Nobel Symposium, Number 84*, ed. S. Bengtson (New York: Columbia University Press, 1994), 3.

16.   S. J. Gould, "'What Is Life?' as a Problem in History," in *What Is Life? The Next 50 Years: Speculations on the Future of Biology*, ed. Michael P. Murphy and Luke A. J. O'Neil (Cambridge: Cambridge University Press, 1995), 36, 38.

17.   S. J. Gould, *The Structure of Evolutionary Theory* (Cambridge, MA: Harvard University Press, 2002), 1339.

18.   B. Russell, *A History of Western Philosophy* (New York: Simon & Schuster, 1945), 659.

19.   Russell, *History of Western Philosophy*, 665–66.

20.   Russell, *History of Western Philosophy*, 670.

21.   Russell, *History of Western Philosophy*, 668, 670.

22.   Russell, *History of Western Philosophy*, 671.

23.   Russell, *History of Western Philosophy*, 672.

5. EXPANDING EVOLUTION

1.   I. Tëmkin and N. Eldredge, "Phylogenetics and Material Cultural Evolution," *Current Anthropology* 48 (2007): 146–54.

8. STARDUST MEMORIES

1.   E. S. Vrba, "Turnover-Pulses, the Red Queen, and Related Topics," *American Journal of Science* 293A (1993): 418–52.

2.   C. Darwin, *On the Origin of Species* (Cambridge, MA: Harvard University Press, 1859), p. 529.

3.   P. W. Goodwin and E. J. Anderson, "Punctuated Aggradational Cycles: A General Hypothesis of Episodic Stratigraphic Accumulation," *Journal of Geology* 93 (1985): 515–33.

4.   A. L. Melott et al., "Did a Gamma-ray Burst Initiate the Late Ordovician Mass Extinction?," *International Journal of Astrobiology* 3 (2004): 55–61.

5.   A. L. Melott, F. Marinho, and L. Paulucci, "Muon Radiation Dose and Marine Megafauna Extinction at the End-Pliocene Supernova," *Astrobiology* 19 (2019): 825–30.

9. IS ETERNAL SEX NECESSARY?

1. Both block quotations are in S. J. Gould, "Is a New and General Theory of Evolution Emerging?," *Paleobiology* 6 (1980): 119-30. In the first, Gould quotes from St. G. J. Mivart, *On the Genesis of Species* (London: Macmillan, 1871), 228-29.
2. Gould, "Is a New and General Theory of Evolution Emerging?," 129.
3. See S. Wright, "Evolution in Mendelian Populations," *Genetics* 16 (1931): 97-159; S. Wright, "The Roles of Mutation Inbreeding, Crossbreeding and Selection in Evolution," *Proceedings of the Sixth International Congress of Genetics* 1 (1932): 356-66; T. Dobzhansky, *Genetics and the Origin of Species*, 3rd ed. (New York: Columbia University Press, 1953); G. G. Simpson, *Tempo and Mode in Evolution* (New York: Columbia University Press, 1944); E. Mayr, *Animal Species and Evolution* (Cambridge, MA: Harvard University Press, 1963).
4. S. J. Gould and N. Eldredge, "Punctuated Equilibria: The Tempo and Mode of Evolution Reconsidered," *Paleobiology* 3 (1977): 115-51.
5. N. Eldredge, " Alternative Approaches to Evolutionary Theory," *Bulletin of the Carnegie Museum of Natural History* 13 (1979): 7-19.
6. E. S. Vrba, "Evolution, Species and Fossils: How Does Life Evolve?," *South African Journal of Science* 61 (1980): 61-84.
7. S. J. Gould, *The Structure of Evolutionary Theory* (Cambridge, MA: Harvard University Press, 2002), 1220-21.

10. DARWIN IN THE GALÁPAGOS

1. N. Barlow, ed., "Darwin's Ornithological Notes," *Bulletin of the British Museum (Natural History) Historical Series* 2 (1963): 33-278.

11. OF CULTURAL NATIONALISM, HAMLET, AND THE *CLOACA UNIVERSALIS*

1. T. C. Chamberlain, "Review of 'Life of James Hall, Geologist and Paleontologist, 1811-1898' by J. M. Clarke," *Journal of Geology* 30 (1922): 175.
2. J. M. Clarke, *James Hall of Albany, Geologist and Palaeontologist, 1811-1898* (Albany, NY: S. C. Bishop, 1925).
3. R. H. Silliman, "The Hamlet Affair: Charles Lyell and the North Americans," *Isis* 86 (1995): 550.
4. Clarke, *James Hall*, 108.

12. WHEN IS A RAPTOR A PARROT?

1. A. Suh et al., "Mesozoic Retroposons Reveal Parrots as the Closest Living Relatives of Passerine Birds," *Nature Communications* 2 (2011): 443; S. J. Hackett et al., "A Phylogenomic Study of Birds Reveals Their Evolutionary History," *Science* 320 (2008): 1763-68. An interesting pop-science take is by the blogger called Grrlscientist, "Jumping Genes

Reveal Birds and Their Sex Chromosomes Evolved Together," https://www.theguardian.com/science/punctuated-equilibrium/2011/oct/24/2.

2.   S. J. Gould and R. C. Lewontin, "The Spandrels of San Marco and the Panglossian Paradigm: A Critique of the Adaptationist Programme," *Proceedings of the Royal Society of London, Series B* 205 (1979): 581–98.

3.   G. C. Williams, *Adaptation and Natural Selection* (Princeton, NJ: Princeton University Press, 1966), 11–12.

4.   G. C. Williams and R. M. Nesse, "The Dawn of Darwinian Medicine," *Quarterly Review of Biology* 66 (1991): 1–22; R. M. Nesse and G. C. Williams, *Why We Get Sick: The New Science of Darwinian Medicine* (New York: Knopf Doubleday, 1996).

5.   S. J. Gould and E. S. Vrba, "Exaptation—a Missing Term in the Science of Form," *Paleobiology* 8 (1982): 7.

References

Ager, Derek V. *The New Catastrophism*. Cambridge: Cambridge University Press, 1993.

Baker, Malcolm P., Brendan Bradley, and Jeffrey A. Wurgler. "Benchmarks as Limits to Arbitrage: Understanding the Low-Volatility Anomaly." *Financial Analysts Journal* 67 (2011): 40–54.

Barlow, Nora, ed. "Darwin's Ornithological Notes." *Bulletin of the British Museum (Natural History) Historical Series* 2 (1963): 33–278.

Clarke, John Mason. *James Hall of Albany, Geologist and Palaeontologist, 1811–1898*. Albany, NY: S. C. Bishop, 1925.

Cuvier, Georges. *Essay on the Theory of the Earth*. 5th ed. Trans. R. Jameson. Edinburgh: W. Blackwood, 1827.

Darwin, Charles. *Charles Darwin's Notebooks, 1836–1844: Geology, Transmutation of Species, Metaphysical Enquiries*, ed. P. H. Barrett, P. J Gautrey, S. Herbert, D. Kohn, and S. Smith. Ithaca, NY: Cornell University Press, 1987.

——. *On the Origin of Species*. Cambridge, MA: Harvard University Press, 1964.

——. *Voyage of the Beagle*. New York: Penguin Books, 1989.

Darwin, Erasmus. *Zoonomia, Vol. 1*. London: E. Earle, 1818.

Dawkins, Richard. *The Extended Phenotype*. San Francisco: W. H. Freeman, 1982.

——. *The Selfish Gene*. Oxford: Oxford University Press, 1976.

DeGregori, James, and Niles Eldredge. "Parallel Causation in Oncogenic and Anthropogenic Degradation and Extinction." *Biological Theory* 15 (2020): 12–24.

Dunbar, Carl O. "Antennae in *Olenellus getzi* n. sp." *American Journal of Science* 5: 303–8.

Eldredge, Niles. "Alternative Approaches to Evolutionary Theory." *Bulletin of the Carnegie Museum of Natural History* 13 (1979): 7–19.

——. "Differential Evolutionary Rates." *Paleobiology* 2 (1976): 174–77.

——. *Eternal Ephemera*. New York: Columbia University Press, 2015.

——. *The Pattern of Evolution*. New York: W. H. Freeman, 1999.

Eldredge, Niles, and Joel Cracraft. *Phylogenetic Patterns and the Evolutionary Process*. New York: Columbia University Press, 1980.

Eldredge, Niles, and Stephen Jay Gould. "Punctuated Equilibria: An Alternative to Phyletic Gradualism." In *Models in Paleobiology*, ed. T. J. M. Schopf, 82–115. San Francisco: Freeman, Cooper and Company, 1972.

Eldredge, Niles, and Marjorie Grene. *Interactions: The Biological Context of Social Systems.* New York: Columbia University Press, 1992.

Eldredge, N., J. Thompson, P. Brakefield, S. Gavrilets, D. Jablonski, J. Jackson, R. Lenski, B. S. Lieberman, M. McPeek, and W. Miller III. "The Dynamics of Evolutionary Stasis." *Paleobiology* 31 (2005): 133–45.

Goodwin, Peter W., and E. J. Anderson. "Punctuated Aggradational Cycles: A General Hypothesis of Episodic Stratigraphic Accumulation." *Journal of Geology* 93 (1985): 515–33.

Gould, Stephen Jay. "Change in Variance: A New Slant on Progress and Directionality in Evolution." *Journal of Paleontology* 62 (1988): 319–29.

—. "Entropic Homogeneity Isn't Why No One Hits .400 Anymore." *Discover*, August 1986, 60–66.

—. "Eternal Metaphors of Palaeontology." In *Patterns of Evolution*, ed. Anthony Hallam, 1–26. Amsterdam: Elsevier, 1977.

—. *Ever Since Darwin.* New York: W. W. Norton, 1977.

—. *Full House.* New York: Henry Holt, 1996.

—. "Introduction: The Coherence of History." In *Early Life on Earth: Nobel Symposium, Number 84*, ed. S. Bengtson, 1–8. New York: Columbia University Press.

—. "The Molluscan Fauna of an Unusual Bermudian Pond: A Natural Experiment in Form and Composition." *Breviora* 308 (1968): 1–13.

—. *The Structure of Evolutionary Theory.* Cambridge, MA: Harvard University Press, 2002.

—. *Time's Arrow, Time's Cycle.* Cambridge, MA: Harvard University Press, 1987.

—. "'What Is Life?' as a Problem in History." In *What Is Life? The Next 50 Years: Speculations on the Future of Biology*, ed. Michael P. Murphy and Luke A. J. O'Neil, 25–39. Cambridge: Cambridge University Press, 1995.

—. *Wonderful Life.* New York: W. W. Norton, 1989.

Gould, Stephen Jay, and Niles Eldredge. "Punctuated Equilibria: The Tempo and Mode of Evolution Reconsidered." *Paleobiology* 3 (1977): 115–51.

Gould, Stephen Jay, Norman L. Gilinsky, and Rebecca Z. German. "Asymmetry of Lineages and the Direction of Evolutionary Time." *Science* 236 (1987): 1437–41.

Gould, Stephen Jay, and Richard C. Lewontin. "The Spandrels of San Marco and the Panglossian Paradigm: A Critique of the Adaptationist Programme." *Proceedings of the Royal Society of London, Series B* 205 (1979): 581–98.

Gould, Stephen Jay, David M. Raup, J. John Sepkoski Jr., Thomas J. M. Schopf, and Daniel S. Simberloff. "The Shape of Evolution: A Comparison of Real and Random Clades." *Paleobiology* 3 (1977): 23–40.

Gould, Stephen Jay, and Elisabeth S. Vrba. "Exaptation—a Missing Term in the Science of Form." *Paleobiology* 8 (1982): 4–15.

Hackett, S. J., R. T. Kimball, S. Reddy, R. C. K. Bowie, E. L. Braun, M. J. Braun, J. L. Chojnowski, W. A. Cox, K.-L. Han, J. Harshman, C. J. Huddleston, B. D. Marks, K. J. Miglia, W. S. Moore, F. H. Sheldon, D. W. Steadman, C. C. Witt, and T. Yuri. "A Phylogenomic Study of Birds Reveals Their Evolutionary History." *Science* 320 (2008): 1763–68.

Hall, James. *Geology of New York.* Albany, NY: Carroll and Cook, 1836–41.

—. *Natural History of New York: Paleontology.* Albany, NY: C. Van Benthuysen, 1847–94.

Hall, James, and John M. Clarke. *Geological Survey of the State of New York, Palaeontology, Volume 7, Text and Plates, Trilobites and Other Crustacea*. Albany, NY: C. Van Benthuysen & Sons, 1888.

Hennig, Willi. *Phylogenetic Systematics*. Champaign: University of Illinois Press, 1966

Kant, Immanuel. 1781. *Critique of Pure Reason*. Cambridge, MA: Houghton Mifflin, 1931.

Lamarck, Jean-Baptiste D. M. de. *Philosophie zoologique, ou Exposition des considerations relatives a l'histoire naturelle des animaux, etc., Vol. 1*. Paris: Dentu, 1809.

Levins, Richard, and Richard Lewontin. *The Dialectical Biologist*. Cambridge, MA: Harvard University Press, 1987.

Lieberman, Bruce, and Adrian Melott. "Declining Volatility, a General Property of Disparate Systems: From Fossils, to Stocks, to the Stars." *Palaeontology* 56 (2013): 1297–1304.

Lyell, Charles. 1832. *Principles of Geology, Volume 2*. Chicago: University of Chicago Press, 1991.

——. *Travels in America, with Geological Observations on the United States, Canada and Nova Scotia*. London: John Murray, 1845.

Mayor, Adrienne. *Fossil Legends of the First Americans*. Princeton, NJ: Princeton University Press, 2005.

Melott, A. L., B. S. Lieberman, C. M. Laird, L. D. Martin, M. V. Medvedev, B. C. Thomas, J. K. Cannizo, N. Gehrels, and C. H. Jackman. "Did a Gamma-ray Burst Initiate the Late Ordovician Mass Extinction?" *International Journal of Astrobiology* 3 (2004): 55–61.

Melott, A. L., F. Marinho, and L. Paulucci. "Muon Radiation Dose and Marine Megafauna Extinction at the End-Pliocene Supernova." *Astrobiology* 19 (2019): 825–30.

Morris, Simon Conway. *The Crucible of Creation*. New York: Oxford University Press, 1998.

——. *Life's Solution: Inevitable Humans in a Lonely Universe*. Cambridge: Cambridge University Press, 2003.

Morris, S. Conway, and S. J. Gould. "Showdown on the Burgess Shale." *Natural History Magazine* 107, no. 10 (1998): 48–55.

Nesse, Randolph M., and George C. Williams, *Why We Get Sick: The New Science of Darwinian Medicine*. New York: Knopf Doubleday, 1996.

Newell, N. "Crises in the History of Life." *Scientific American* 208 (1963): 76–95.

Raup, D. M., and G. E. Boyajian. "Patterns of Generic Extinction in the Fossil Record." *Paleobiology* 14 (1988): 109–25.

Raup, D. M., and S. J. Gould. "Stochastic Simulation and Evolution of Morphology: Towards a Nomothetic Paleontology." *Systematic Biology* 23 (1974): 305–22.

Raup, D. M., S. J. Gould, T. J. M. Schopf, and D. S. Simberloff. "Stochastic Models of Phylogeny and the Evolution of Diversity." *Journal of Geology* 81 (1973): 525–42.

Russell, Bertrand. *A History of Western Philosophy*. New York: Simon & Schuster, 1945.

Silliman, R. H. "The Hamlet Affair: Charles Lyell and the North Americans." *Isis* 86 (1995): 541–61.

Silverstein, A. "An Aristotelian Resolution of the Idiographic Versus Nomothetic Tension." *American Psychologist* 43 (1988): 425–30.

Stanley, Steven M. *Macroevolution*. Baltimore: Johns Hopkins University Press, 1979.

Suh, A., M. Paus, M. Kiefmann, G. Churakov, F. A. Franke, J. Brosius, J. O. Kriegs, and J. Schmitz. "Mesozoic Retroposons Reveal Parrots as the Closest Living Relatives of Passerine Birds." *Nature Communications* 2 (2011): 443.

Tëmkin, I., and N. Eldredge. "Phylogenetics and Material Cultural Evolution." *Current Anthropology* 48 (2007): 146–54.

Vrba, Elisabeth S. "Evolution, Species and Fossils: How Does Life Evolve?" *South African Journal of Science* 61 (1980): 61–84.

——. "Turnover-Pulses, the Red Queen, and Related Topics." *American Journal of Science* 293A (1993): 418–52.

Wiley, Edward O. *Phylogenetics: The Theory and Practice of Phylogenetic Systematics.* New York: John Wiley & Sons, 1981.

Wiley, Edward O., and Bruce S. Lieberman. *Phylogenetics: The Theory and Practice of Phylogenetic Systematics.* 2nd ed. Hoboken, NJ: Wiley-Blackwell, 2011.

Williams, George C. *Adaptation and Natural Selection.* Princeton, NJ: Princeton University Press, 1966.

Williams, G. C., and R. M. Nesse. "The Dawn of Darwinian Medicine." *Quarterly Review of Biology* 66 (1991): 1–22.

Windelband, W. "History and Natural Science." *History & Theory* 19 (1988): 169–85.

Index

Milton Keynes UK
Ingram Content Group UK Ltd.
UKHW041402011024
448963UK00007BA/13/J

9 780231 208109